CAR TROUBLES

T0264804

Car Troubles
Critical Studies of Automobility and Auto-Mobility

JIM CONLEY
Trent Univeristy, Canada
and
ARLENE TIGAR McLAREN
Simon Fraser University, Canada

Routledge
Taylor & Francis Group

LONDON AND NEW YORK

First published 2009 by Ashgate Publishing

2 Park Square, Milton Park, Abingdon, Oxon OX14 4RN
711 Third Avenue, New York, NY 10017, USA

Routledge is an imprint of the Taylor & Francis Group, an informa business

First issued in paperback 2016

British Library Cataloguing in Publication Data
Car troubles : critical studies of automobility. --
  (Transport and society)
  1. Transportation, Automotive--Social aspects.
  2. Automobiles--Social aspects.
  I. Series II. Conley, Jim. III. McLaren, Arlene Tigar.
  303.4'832-dc22

Library of Congress Cataloging-in-Publication Data
Car troubles : critical studies of automobility / edited by Jim Conley and Arlene Tigar McLaren.
    p. cm. -- (Transport and society)
  Includes bibliographical references and index.
  ISBN 978-0-7546-7772-7
1. Automobiles--Social aspects--History. 2. Transportation and state--History. I. Conley, Jim, 1951- II. McLaren, Arlene Tigar.

  HE5611.C26 2009
  388.3'420973--dc22

                                                                    2009007621

ISBN 978-0-7546-7772-7 (hbk)
ISBN 978-1-138-25534-0 (pbk)

Contents

PART 3: INEVITABLE AUTOMOBILITY?

PART 4: BEYOND THE CAR

List of Figures

List of Tables

Contributors

Catherine Bean, School of Geography, Geology and Environmental Science, University of Auckland

Damian Collins, Department of Earth and Atmospheric Sciences, University of Alberta

Jim Conley, Department of Sociology, Trent University

Kingsley Dennis, Department of Sociology, Lancaster University

Catharine Genovese, Department of History, Simon Fraser University

Jason Henderson, Department of Geography, San Francisco State University

Robin Kearns, School of Geography, Geology and Environmental Science, University of Auckland

Todd Litman, Executive Director, Victoria Transport Policy Institute

David MacGregor, Department of Sociology, University of Western Ontario

Arlene Tigar McLaren, Department of Sociology and Anthropology, Simon Fraser University

Fiona McLean, Coordinator, Car Fuel Efficiency Campaign, Friends of the Earth Europe, Brussels

George Martin, Department of Sociology, Montclair State University

Derek Simons, School of Communication, Simon Fraser University

Dennis Soron, Department of Sociology, Brock University

Patricia Tomic, Department of Sociology, University of British Columbia, Okanagan

Ricardo Trumper, Department of Sociology, University of British Columbia, Okanagan

John Urry, Department of Sociology, Lancaster University

Jameson M. Wetmore, School of Human Evolution and Social Change, Arizona
State University

Acknowledgements

We thank the Social Sciences and Humanities Research Council (SSHRC) of Canada for supporting the Traffic Safety Project from its inception as a Research Development Initiative in 2004 to the October 2006 symposium entitled "autoConsequences: Automobilization and its Social Implications" held in Vancouver, British Columbia. Simon Fraser University, including the Dean of Graduate Studies and the Dean of Arts, financially contributed to the symposium. We thank John Urry for his interest in the symposium and for serving as the keynote speaker. Colleagues and staff in the Department of Sociology and Anthropology at SFU, and particularly the Chair Jane Pulkingham, provided consistent support and encouragement. Arlene is grateful for Jim Conley's early and continued involvement and the project would not have been possible without the enthusiasm, interest and hard work of graduate students Sylvia Parusel, Lucie Vallières, and Stephen Carley who assisted with the research and the symposium. A special thanks to Sylvia for her unfailingly meticulous assistance and to Margaret Manery for her expert help in finishing the manuscript. We thank Jean Wilson and Melissa Pitts for their encouragement and the anonymous reviewers for their insightful and constructive criticism on earlier versions of the text. We are grateful to Margaret Grieco, series editor at Ashgate, for her interest in the text and facilitation of its publication. We appreciate the work and guidance of Val Rose, Jude Chillman and Maureen Mansell-Ward in producing the book.

Arlene thanks friends in Victoria, Vancouver, and London for their love, interest and support as well as her parents, Margaret and Bud Tigar, who first introduced her to the joys of the car at a time when its contradictions were less evident. She is forever grateful to her strongest supporters and critics, Angus and Jesse.

Jim wishes to thank Arlene for organizing the symposium, and inviting him to co-edit this book. He also thanks Jay Owen of Volvo Cars of Canada for so promptly giving permission to use one of their advertisements in his chapter, and Sam Petherbridge of Sharpe Blackmore Euro RSCG for supplying the image. He is grateful to the Department of Sociology at Trent University for providing the flexibility to offer a course in Sociology of the Automobile, and to the students in that course. His late parents, Mike and Millie Conley, showed that it was possible to live without a car in mid-twentieth century Edmonton, thus helping to put him in a position of optimal marginality for a sociologist of automobility. Shelley, Patrick, and Gregory cannot be thanked enough for their love and support.

Introduction

Jim Conley and Arlene Tigar McLaren

> I suppose you could get all worked up about the wretched excess of the whole thing
> – but that would be just small minded, wouldn't it? We'd all have one, or something
> similar, if we could, wouldn't we? (English 2008)

So concludes automotive journalist Bob English's road test of the 2008 Lexus LS600hL Luxury Hybrid Sedan, a car that boasts a base price of $125,400, both a 5.0 litre V8 engine and electric motors, "exterior styling [that] combines that nice touch of elegance and arrogance," and a "stunning" level of luxury. But would we all want one? This volume challenges the assumption that desire for such an automobile is uncontroversial and natural. At a time when cars are increasingly reliable mechanically, *Car Troubles* is a multidisciplinary exploration of other kinds of trouble associated with the growing global reach of the automobile.

Automobility and Auto-Mobility

The car in modern societies entails both experience, which we refer to as 'auto-mobility' and a system of which it is a part – automobility. For most people the experience of car travel includes the act of driving a car or feelings towards it and implies autonomous, flexible and speedy travel. Yet automobility is a complex and expanding system that makes driving a car possible and even necessary (Paterson 2007). In Urry's definition, automobility is "the self-organizing, self-generating, non-linear world-wide system of cars, car-drivers, roads, petroleum supplies, and many novel objects, technologies, and signs" (2004, 27).

Paradoxes abound with both auto-mobility and automobility. Anyone who has felt the thrill of speeding along an open road, the irritation of getting stuck in gridlock and the fear of crossing a road as cars barrel towards them knows the automobile as both liberating and constraining. Even in its self-propelled forms such as cycling or walking, auto-mobility is virtually impossible because paths, roads and so on are required (Böhm et al. 2006a; Paterson 2007). In the case of the motor vehicle, the contradictions are more apparent. Autonomous motorized travel depends on the system of automobility to sustain and promote it, while the latter cannot exist without people who need or want auto-mobility. Neither can be reduced to the other and their relation of dependence is the source of consequences and contradictions explored in this volume. Together they encompass an array of interlocking social, cultural, technical, political and economic forces.

By examining the relation between automobility and auto-mobility, this book does not simply highlight the car's negative consequences. Instead, it problematizes, or troubles the car by considering how and why it is so deeply embedded in social life. Auto dependence for mobility can seem inevitable. Only in crises (oil shortages, failures of infrastructure, chronic congestion, rising fuel prices, economic downturns and so on) does the automobility system that gives rise to and sustains the car become an object of concerted public and media attention.

Rather than assuming essential characteristics of auto-mobility (e.g., that it is inherently an expression of freedom), or that the system of automobility is all-powerful, we suggest it is important to socialize the analysis. In pursing this objective, this volume takes on a range of approaches that includes social theory, specific case studies and/or policy analysis in diverse disciplines (for example, sociology, geography, social and technical studies, environmental science, history, economics, transport policy). Given its iconic status in modern production and consumption and its critical relationship to urban design, the automobile is a particularly fruitful object for multi- and interdisciplinary exploration. New trends in scholarship such as the 'mobility turn' in various disciplines (Hannam et al. 2006), work that brings together technical and social studies (Beckmann 2004; Norton 2008), and governmentality and cultural studies that investigate subjectivity (Paterson 2007; Packer 2008), have helped to invigorate the study of automobility.

While interdisciplinarity is crucial for understanding automobility, it is challenging. The theoretical and methodological conventions of one discipline can appear bewildering and unsatisfactory to another. Yet the complexity of automobility, we suggest, requires nothing less than the multi-faceted lens of various disciplines. The traditional separation of technical and social studies, for example, does not suffice for adequate theoretical explanations and strategies for change. Working across disciplines allows for greater insight into how automobility is entangled with material and social life globally and in specific cultural contexts (for example Derek Simons' chapter brings together insights from Western art history, science and technology).

While this volume is part of a growing chorus of criticism, it places itself in the intersection of the undoubted appeals of auto-mobility and the equally undoubted harms of automobility. Understanding the former is necessary for successful strategies to address the latter. The volume explores the reasons for a critical view of automobility and for its worldwide dominance over other modes of land transport. In this introduction and the book as a whole, we emphasize the recurring car troubles of: environmental unsustainability; economic wastefulness; death and injury; and social dislocations, inequities and exclusions. We also address debates about the economic and cultural power of automobility – why it prevails as a local and global system – and how change may come about in a complex system and politics of automobility.

The Critique of Automobility: Theoretically Troubling the Car

Almost since its inception, some people have found the car troubling. At the beginning of the automobile age in the US, when cars were playthings of the rich, driven too fast, frightening horses and killing pedestrians, class exclusion fuelled critiques (McShane 1995; Norton 2008). The advent of Fordism and the mass production of cheaper vehicles defused the class critique. Crucial to its success in sparsely settled countries such as the United States and Canada, with a less unequal distribution of income than elsewhere (Flink 1988), the automobile originally had much to recommend it relative to the alternatives. Even so, into the 1930s and beyond in the US, cars continued to be disquieting: they allowed youth to circumvent adult controls and social norms; upset established status relations; disrupted and destroyed communities and depopulated small towns; and in cities produced traffic congestion and "the parking problem" (Flink 1988, 151–52; Lynd and Lynd 1929; 1937). After World War II, when mass motorization became highly advanced in North America and Western Europe, a sustained public critique of automobility emerged that focused on American automakers for their shoddy products, manipulation of consumers, and neglect of safety (Keats 1958; Nader 1965). In the 1990s, and closely connected to environmentalism, criticism of automobility accelerated in both popular (e.g., Alvord 2000) and academic publications (Miller 2001a; Featherstone et al. 2005; Böhm et al. 2006a; Paterson 2007).

Recent scholarly interest in the automobile is due to a host of reasons that include its increased production and consumption in developed and developing countries and the gnawing sense of the growing risks and troubles associated with automobile-dominated transportation. The advantages of auto-mobility to individuals and society have become curses in large part because mass motorization produces economic, social, health, environmental, and institutional externalities. That is, the benefits of automobile ownership and use largely accrue to individuals, while the costs of that consumption in the aggregate are borne by society at large (Miller 2001b), and often differentially by specific groups in society (see below). Though drivers or car-owners pay the private costs of automobile use (gas, oil, maintenance, the price of the vehicle itself, insurance), and reap some private benefits (mobility, status, pleasure), they do not pay directly the full costs of the automobility system (including enormous public investment in and maintenance of infrastructure such as roads, bridges and parking spaces, to say nothing of pollution, urban sprawl, and death and injury from traffic collisions) (Freund and Martin 1993). Externalities include in the case of risk and safety, for example, vehicle designs that protect drivers and passengers to the detriment of pedestrians and cyclists (or occupants of smaller vehicles – when people buy big SUVs because they feel safer in them despite the increased risk of rollovers – Bradsher 2002; Insurance Institute for Highway Safety (IIHS) 2005). Other externalities include institutional entailments necessary for automobile transportation (e.g., the police, courts, license bureaus and so on that regulate cars and drivers) – returning us once again to automobility as a vast interlocking system.

Yet, those who drive are not separate from externalities: as taxpayers they pay for public (but often hidden) costs of automobility and are non-drivers, if they walk or cycle. Further, the aggregate use of motor vehicles turns their consequences back onto the individual user, as congestion, pollution, risk of injury and so on. As Beck (1992) argues, risks in society often have a boomerang effect: those who contributed to the ill effects also experience them and therefore may be motivated to address the problems. Yet, the question remains of how individuals and groups in society will effect change to reduce the risks.

As C.W. Mills (1959) argued, to advance social change it is critical to demonstrate the links between private troubles and public issues. Yet for many people the car is not a private trouble; it is a solution to private concerns, such as access to jobs, housing, recreation and social status. When the private troubles of automobility have become public issues, the approach usually taken seeks to improve automobility: making cars safer for their occupants; removing dangerous drivers from the roads; and creating technical changes that leave consumption patterns unchanged. In contrast, this volume raises fundamental questions about the viability of the automobility system.

Environmental Unsustainability

Growing worries about the natural environment account for much of the recent upsurge in criticisms of automobile transportation. Environmentalist critics condemn cars for their contributions to global climate change and air and water pollution, their depletion of non-renewable resources, especially oil, and their land use impacts (e.g., Böhm et al. 2006b; Newman and Kenworthy 1999; Paterson 2007). The environmental impacts of automobility form the backdrop to several chapters in this volume (Jason Henderson; Todd Litman; George Martin; Kingsley Dennis and John Urry). Several other chapters more directly address the automobile's complex and contradictory relation to urban life, the suburbs and cultural images of nature. Both Jim Conley and Fiona McLean document how automobile advertising appropriates images of nature while neglecting the impact of the car on the environment. The automobile has provided a technology for escaping the conditions and consequences of its own production to a romantically conceived nature that, whether it is a pure, pastoral Arcadia, or a harsh, dangerous place of adventure, is accessed and mastered by the same technology that destroys it (Williamson 1978). Paradoxically, growing environmental concerns for nature are in part a product of automobility (Flink 1988; Sachs (1992) including its encroachment as systems of roads and suburbs develop. In light of such contradictions, McLean raises the question of how advertising messages might shift as the industry responds to growing concerns about the contribution of SUVs to climate change.

In contrast to Conley's and McLean's cultural analyses, Dennis Soron criticizes environmentalist anti-car politics for neglecting material constraints on individual choice and placing too much responsibility on individual consumers for the evils of auto dominance. In arguing that the consumption of cars is compulsory in automobilized societies, Soron highlights a debate over the priority of material and cultural analyses that runs through this volume.

Economic Waste

To the extent that the production and consumption of automobiles has been an engine of economic growth and a source of well-paying jobs (Paterson 2007), and that possession of an auto industry has been a sign of modernity and source of national pride (Edensor 2004; Garvey 2001; Koshar 2004; Sachs 1992), it is counter-intuitive to base the critique of automobility on economic wastefulness. As Catharine Genovese's chapter on hot rodders in 1950s Vancouver Canada illustrates, the vast and proliferating auto system afforded a range of mainstream and alternative economic opportunities. Indeed, the very success of automobility as a pervasive system makes it difficult to calculate economic costs and benefits – for example, how policing, health care and other services subsidize automobile travel (cf. Miller 2001a). Yet, using conventional economic reasoning, Litman provides grounds for considering the automobile, in particular as a status symbol, to be economically wasteful. To the degree that auto-mobility and other forms of mobility are prestige goods, and therefore zero-sum, then auto consumption is socially wasteful. By inducing more expenditure on cars than would otherwise occur if it were not a source of status, auto-mobility is an economic trap.

Other chapters also suggest that automobility is economically wasteful (Henderson; Ricardo Trumper and Patricia Tomic). Martin's analysis of the expansion of automobility in less developed countries implies that one of the 'advantages of backwardness', namely the lack of sunk costs in automobile infrastructure, provides the opportunity to spend transportation funds more wisely than more developed countries have.[1] If these infrastructure costs (roads, bridges, pipelines, oil sands projects) needed to support automobility are also economically wasteful at the societal level, recent failures of the latter (e.g., bridge collapses in Laval, Quebec in 2006 and Minneapolis, Minnesota in 2007) catastrophically illustrate a collision between economic and political logics. More political capital is often to be gained by opening new infrastructure projects, such as freeways, that contribute to urban sprawl and auto dependence, than in repairing and renewing existing infrastructure, and developing better alternatives. At the beginning of 2009, as governments provide bailouts to automakers and consider infrastructure spending to counter a deepening recession, it seems doubtful that, in the short

---

1   In Canada in 2005, 40 percent of the dollar value of government-owned infrastructure consisted of roads and bridges (Roy 2007), a figure that is probably representative of other developed countries.

term at least, a more critical systemic approach to automobility will appear in politics as it has in scholarship. Such an approach suggests that what seem to be simply economic and engineering problems would benefit from a multidisciplinary approach, a lesson also conveyed by a third basis for the critique of automobility.

Carnage: Death and Injury

Automobility raises public health issues that range from the health effects of urban sprawl such as the loss of walking space and opportunities for physical exercise in day-to-day life (Freund and Martin 1993) to evidence that automobile pollution has contributed to respiratory ailments and premature death. Public health strategies, for example, include increased land use mix for promoting walking and diminishing car use to reduce problems of obesity (Frank et al. 2004).

Most tragically, however, the public health impact of automobility appears in the form of death and injury in motor vehicle collisions. Approximately 1.2 million people die each year in motor vehicle collisions worldwide, and countless others are injured (Peden et al. 2004). Jain (2004, 61) asks: "What are we to make of this susceptibility to injury and the lack of recognition of this major, but potentially rectifiable public health issue?" While an environmental movement has argued effectively that cars pollute, the safety movement has stalled rather than surged forward. It has maintained a strategy of improving the car, not seeking alternatives to it. Safety only intermittently appears on the radar of public concerns; it remains primarily as a routinized technological, legal or educational issue remote from central political, cultural and social agendas, despite the fact that it is a public health problem, and anything but the accidental result of automobility.

Several chapters explore the cultural, social and technical dimensions of automobile sacrifice and safety. Emphasizing speeding projectiles – artillery shells, locomotives, automobiles – hurtling through concrete, Derek Simons roots the death and mayhem of automobility in a 'will-to-sacrifice' located historically in Western culture. While agreeing that a sacrificial urge may be present, David MacGregor shows that states vary in their willingness to regulate individual behaviour or the automobile industry. Even as a safety race is emerging amongst automakers in a new cultural age of the automobile, MacGregor claims, the state must be the major promoter of auto safety. Jameson M. Wetmore's historical analysis of the air bag demonstrates that safety innovations have multiple sources, including the state, the car industry and other sectors of society. He argues that in the United States safety advocates were most successful when instead of one-sidedly pursuing either 'technical fixes' (regulating industry) or 'social fixes' (regulating driver behaviour) they distributed responsibility by promoting both simultaneously. Wetmore's analysis exemplifies the science and technology studies maxim that technical artefacts and social relationships cannot be considered separately. Similarly, Genovese shows how hot rodders, in facing public disquiet and disapproval, sought respectability by adopting safety strategies of both regulating 'outlaw' driver behaviour and improving the technical performance of vehicles. In contrast, in their study of walking school buses in Auckland, New Zealand, Damian Collins, Catherine Bean and Robin Kearns find that adults make child pedestrians responsible for collisions by emphasizing the need to discipline and regulate their behaviour, not that of drivers. Their research supports other literature that considers how traffic safety discourse privileges car-based transportation, ensuring that pedestrians do not impede automobility (Jain 2004; Norton 2008; Vallières 2006).

The debate about safety and the impact of automobility raises the question of how responsibility is allocated and thus of how power is distributed and exercised. How, for example, do neoliberal governments choose to discipline and regulate the 'responsible agents' or support transport alternatives to the car (McLaren 2007)? The allocation of responsibility brings us to the social dependencies and externalities of automobility – the contradiction between the individual benefits of car-driving and its social costs.

Community, Sociability and Inequality

The individual mobility of auto-mobility forms both part of its appeal and a powerful basis for its critique. Critics of the social implications of automobility have focused on its spatial effects, its relation to excessive individualism and its creation of social inequalities and exclusions of specific populations.

Much of the critical literature of automobility's spatial effects has centred on cities, in the form of urban sprawl: the hollowing out of central business districts; the dispersion of places of residence, work, leisure and shopping to far-flung suburbs that are difficult to service effectively with public transportation; and traffic congestion. Jane Jacobs' influential *Death and Life of Great American Cities* (1961), like Lewis Mumford's (1964) *The Highway and the City*, blamed not the automobile per se, but urban planning that reduced diversity and promoted over-dependence on a single mode of transportation for the destruction of the dense face-to-face sociality of urban neighbourhoods.

While sprawl is a contested domain (see Henderson's chapter for a specific case study), it is facilitated by and contributes to the hyperautomobility analyzed by Martin and the self-reproduction of automobility emphasized by Dennis and Urry. Simons' chapter is a reminder of how the automobile has shaped the built, material environment: concrete overpasses, underpasses, on-ramps, bridges, pillars and dividers along which motor vehicles hurtle like bullets through the barrel of a machine gun. His analysis will resonate with anyone who has walked in a wholly auto-dominated environment.

But automobility begets ironies: on the one hand, automobile travel makes possible the maintenance of social networks amongst geographically dispersed people, at the same time that it contributes to that dispersion – another way in which it is reproduced by both creating a problem and solving it (Urry 2004). New communities and connections may be formed through automobility (Sheller and Urry 2000). Cars do not just destroy community; for those who have access to them they also make possible new kinds of community and identity, not rooted in geographical proximity, as Genovese indicates in her chapter on hot rodders. Likewise, even though urban space prioritizes motor vehicles over pedestrians and thereby creates "socially dead 'public' spaces dominated by vehicular traffic and associated externalities," Collins, Bean and Kearns show that routinized walking at the neighbourhood level both accommodates and challenges the subjectivity and social organization associated with automobility. While families appreciated the benefits of auto-mobility in extending networks and opportunities, they found the walking school bus less private and isolating; it helped to create socially vibrant neighbourhoods. And as authors as diverse as Jacobs (1961) and Augé (1995; 2002) have argued, collective modes of travel themselves create such 'non-places' as airport lounges or subway cars that are hardly conducive to sociability, at least in a private, local or intimate sense.

Some scholars celebrate an affinity between auto-mobility and the individual autonomy of liberal political philosophy (Dunn 1998; Lomasky 1997). Cars potentially free people from social regulation in a variety of ways: from direct dependence on other people or organizations and regimentation by their schedules and routes; from face to face interaction with others not of one's choosing such as occurs on railways and mass transit; and from social obligations, especially those associated with civilization and domesticity.

On the other hand, holding out the possibility that we can go where we want, when we want, with whom we want, and at the speed we want, the auto's flexibility creates conditions for the anomic, unlimited desires against which Durkheim (1966) warned. The anomie inherent in car travel is exacerbated by the coerciveness of flexibility and the consequent need to "juggle tiny fragments of time" (Sheller and Urry 2000, 744). As Sheller and Urry (2000; Urry 2004) have noted, the unprecedented individual flexibility and autonomy provided by cars is coercive, because once generalized it comes to be expected – we are forced to be flexible. In providing privacy, cars isolate their occupants and impair their interaction with occupants of different cars (Lupton 1999; Sheller and Urry 2003; Urry 2006). Especially under the conditions of lifestyle differentiation (Gartman 2004), cars are ideal vehicles for egoism, as each driver's wants, frustrated by everyone else trying to achieve theirs, weaken solidarity with non-car users of the road and occupants of different types of vehicle. As Sachs (1992, 176, 177) puts it, when everyone is behind the wheel, "their desires get in the way of other desires" and "everyone is stealing everyone else's precious time; annoyance reigns everywhere, and rage."

In Martin's terms, hyperautomobility and hyperindividualism appear to go together with the increased use of the privatized, individualized automobile. In their chapter on road-building in Chile, Trumper and Tomic argue that auto ownership and use is promoted as part of a neoliberal modernization project in which the "car [is] a hybrid that changes drivers' views of the world, making them more consumerist, individualistic, and concerned with the pursuit of self-determined private purposes."

Critics of automobility have not just blamed it for weakened communities and excessive individualism; they have also attacked its implications for a variety of social inequities and exclusions of class, racial and ethnic group, gender, age and ability. The benefits of automobile ownership and use generally are unequally distributed; some segments of the population disproportionately bear its costs, while others disproportionately reap its benefits.

As in much else, the externalities of automobility are borne especially by the poor, and by the racially marginalized (Freund and Martin 1993). Those who cannot afford to buy an automobile have fewer job opportunities, particularly as a result of urban sprawl. With white flight in Atlanta, Georgia, for example, inner-city blacks without a car are unable to gain access to suburban jobs (see Henderson). When freeways cut through their neighbourhoods the poor suffer more automobile pollution, and in particular cases such as Chile, they undergo the added inequity that workers' compulsory pension savings are used to build freeways from which they do not benefit (see Trumper and Tomic). While contrasting the hyperautomobility of the US to the emerging mass motorization in countries such as India (Waldman 2005) and China, Martin's chapter indicates that increased motorization in China – and its emergence as a world-scale consumer and producer of automobiles – has produced urban sprawl, displacement of the poor and migrant workers to urban peripheries, and further social fragmentation.

Complex gender inequalities are also associated with automobility. Since its inception, the automobile has been associated symbolically with masculinity, especially when technical knowledge is involved (Freund and Martin 1993). The masculine bias of North American car culture continues to be reproduced in the way in which some vehicle types have developed stereotyped feminine associations: minivans with 'soccer moms'; 'chick cars' – small sports cars, and small SUVs (or 'cute-utes') – with young women. Yet, changes in family forms, the growing prevalence of multicar, multiple earner households, and perhaps changes in masculinity and femininity themselves contribute to shifts in popular culture to representations of the car as "a highly ambiguous gendered space" (Jain 2005, 187).[2] Several chapters in this volume consider the cultural significance of gender in automobility: in hot rodding, automobile advertisements and walking school buses. The latter study (Collins, Bean and Kearn) also attends to one of the most intractable exclusions that can be directly tied to automobility, which is age, especially for children. Scholars have long recognized the precarious position of children in auto-dominated urban environments as their play spaces are taken over by motor vehicles (Freund and Martin 1993; Lynd and Lynd 1937; Norton 2008).

---

2   Beyond car use and the representations of cars, however, little research has examined how the other facets of the automobility system are gendered (e.g., who designs the cars or plans urban environments, with what conceptions of masculinity and femininity).

The Economic and Cultural Power of Automobility

Auto-mobility continues to be celebrated despite automobility's troubles. Its worldwide expansion suggests that auto-mobility has considerable appeal to large numbers of people. But what is the basis of its appeal and power? Scholarly debate about the power of automobility divides roughly into 'cultural' versus 'materialist' explanations. The debate between them, which is as old as social science itself (see Sahlins 1976), is complex and unlikely to be resolved here. Nonetheless, automobility offers an ideal site for scholarly explorations of materialist and cultural explanations.

In this volume, several chapters highlight materialist explanations of the power of automobility: the physical organization of urban and suburban space, utilitarian motivations for car use within that space, and economic interests that seek to profit from the production, expansion and consumption of cars. Other chapters emphasize cultural explanations: symbolic representations, discursive and ideological formations, social identities and social motivations in relation to the car. Additional chapters suggest, however, that the material and cultural are so intertwined that it is not possible to separate them in explaining auto hegemony.

In his chapter on 'compulsory consumption', Soron forcefully argues the materialist position. He maintains that compulsory automobility persists due to "the cosy and often incestuous relationship between private industry and government." Once a full-fledged system of automobility is established, it is increasingly difficult for individuals to engage in the normal run of daily activities without having use of a private automobile, particularly with the physical constraints created by urban sprawl. Soron contends that regardless of what individuals may feel about cars, they need them for the flexibility they provide in the daily round of activities in highly dispersed locales. Soron's argument helps to explain why, even when they recognize the aggregate problems of automobility, people are constrained in having to rely on cars for instrumental reasons.

Several chapters investigate which groups have the most power for making decisions about the material shape of the environment, particularly the ways that capitalist interests invest economically and politically in mass motorization. The business of automobility is spread widely throughout an auto-industrial complex including auto manufacturers and dealers, petroleum and rubber producers, road builders, and real estate developers, who operate at local, national and global scales. Trumper and Tomic explain the development of automobility in Chile as a neoliberal capitalist accumulation strategy, from which transnational corporations, the political elite and the affluent middle classes benefit, not the lower and poorer classes of Chile. Martin argues that the complex of auto-oil-construction firms, "coupled with the support of governments, is the basis of a powerful bloc in the global economy" that promotes auto hegemony. While agreeing on the power of the auto-industrial complex, Henderson in contrast shows that capitalist interests in Atlanta were by no means united. On the one hand, powerful corporate interests opposed unrestrained automobility; on the other hand, a largely white, middle-class anti-urban cultural ethos that combined "rural idealism, 'family values' and fundamentalist religion" supported secessionist automobility in alliance with real estate and road-building interests. Material interests are not the whole story, as they are intertwined with cultural concerns that cohere around 'race', family and religion.

Without a doubt, automobile manufacturing is a powerful economic force in the more developed countries and increasingly in the less developed countries. As Conley's chapter notes, in the first half of 2005 alone, 29 auto brands were estimated to have spent over 5 billion dollars in US media. Such pervasive and intensive advertising raises the question of the extent to which the auto industry manipulates consumers' desires and needs. Conley disputes the view that advertisers simply manipulate consumers and instead argues that advertisers draw upon a stock of collective representations through which people make sense of their lives within automobilized societies. His semiotic analysis shows how central cultural oppositions such as excitement and safety, and masculine and feminine combine the mundane and magical sides of automobility, with excitement from speed, for example, as a prominent symbolic representation in car advertising. McLean argues in her analysis of SUV advertising that by systematically appropriating cultural representations, the auto industry contributed to the surge in the popularity of the SUV in North America and elsewhere. By the design of images that associate the SUV with glorified natural environments to escape from urban settings, advertisers take advantage of consumers' attachments to symbols and meanings with deep roots in Western culture.

Litman also offers a cultural explanation in arguing that individuals desire automobiles not simply for transportation, but to acquire status. In his analysis of mobility as a prestige good, Litman maintains that cars signify social mobility and social status for individuals or families: what you drive is a key marker of status. He makes the point that the utilitarian argument does not address the prior question of why people were attracted to automobiles, suburbs and so on in the first place.

Beyond being status symbols, cars also create cultural opportunities for the experience of power, speed, and excitement, showing the driver's skill and daring by taking risks and transgressing norms (Freund and Martin 1993; Gitlin 1986), as Genovese shows in her case study of hot rodders in mid-twentieth century Vancouver. This chapter illustrates the notion of 'affordances' (Hannam et al. 2006) and the interpretative flexibility of technical artifacts (Kline and Pinch 1996): in addition to its prosaic, utilitarian transportation functions, the car provides opportunities to do things and have experiences not necessarily anticipated by their makers (although seeing a potential market, they or other capitalists soon move to supply it).

Several chapters that address the issue of safety and automobility combine material and cultural explanations. MacGregor stresses the importance of the state as well as recent technological developments for the emergence of a new cultural age in which 'safety sells'. Wetmore's history of the air bag draws clearly on the science and technology approach, arguing that technical artefacts and social relationships cannot be considered separately. Collins, Bean and Kearns focus on the 'social fix' of walking school buses, but this strategy also has the side-effect of redesigning the material world by reducing the number of cars on the road and thus reshaping urban space. Simons challenges directly the very distinctions between the material/objective and the social/subjective in his interdisciplinary analysis of the material culture of concrete and the projectile economy.

In drawing on complexity theory, Dennis and Urry adopt a highly fluid notion of the material and the cultural in assessing the current state of automobility and projecting its future. They observe that both economies and social life have been 'locked in' to the 'steel-and-petroleum' car as a result of relatively small causes that laid down an irreversible pattern, ensuring the preconditions for automobility's self-expansion over the past century: "supported through a huge economic, social, and technological maelstrom of vested interests, agents, and interrelated flows." While the current structure of the car system is remarkably powerful, stable and unchanging, Dennis and Urry suggest, this complex assemblage is neither socially necessary nor inevitable.

Beyond Critique – Beyond the Car

Any critique of automobility requires going beyond what Henderson in his chapter calls the 'inevitability hypothesis', that is, the unquestioned assumption that the domination of the automobile over all other forms of transportation is inescapable. But once the inevitability of automobility is challenged, what lies beyond the critique of car hegemony? Chapters in this volume address in diverse ways the question of how we get 'beyond the car'.

First, what is to be changed? Is it automobility or automobilities, car culture (singular) or car cultures (plural)? To put it differently, once a society embarks on mass motorization, do certain consequences inevitably follow? Chapters vary in their emphasis on either inherent features of automobility that make it self-reproducing, or on different, particularistic automobilities that depend on specific conjunctions of interests, circumstances, national cultures, local politics and social structures.

Second, who or what are the agents of change? In this volume, the main candidates include consumers, states, corporations, social movements and the system of automobility itself. While consumers potentially have an impact on changing automobility, none of the chapters highlight their significance. To the extent that the problems created by individual auto-mobility are externalities in which the collective costs of automobility are not borne by the individual driver, getting beyond the car is caught in a collective action or free-rider problem that renders individual consumer action an unpromising avenue for change: individuals who use alternative means of transport, and accrue extra costs in doing so (added commuting time, inconvenience, transit fares, loss of status) reduce congestion and thus make car travel less costly. Each cyclist with a 'one less car' sticker, and each subway or bus rider makes it easier for others in cars to reach their destinations, so the cumulative impact may be less than hoped for unless there are changes in urban design that give priority to non-car modes of transportation (Newman and Kenworthy 1999). On the other hand, the worsening position of American automakers, as increasing fuel prices have shifted North American consumers away from SUVs and trucks to smaller, more fuel-efficient vehicles, should caution against prematurely dismissing consumers as a source of change.

Some chapters in the book imply that change can come about primarily through the action of states, and perhaps only anti-capitalist or post-capitalist states, or the state in conjunction with the car industry and other sectors of society. Others emphasize local community action and social movements that consciously attempt to direct change. While social movements rarely achieve all that they strive for and their effects are difficult to estimate (Tarrow 1998), some movements related to automobility have had notable successes, such as those for automotive safety, and against drunk driving. Yet neither have challenged auto hegemony, and the success of the latter has depended upon both powerful state allies and a congenial ideological climate (Reinarman 1988; McCarthy 1994). With the exception of campaigns against freeways in the 1960s, movements that more directly challenge automobility have had more mixed results. For example, environmentalists have achieved some success in limiting automotive emissions, but judging by continued road-building agendas in many countries, anti-car movements such as 'Reclaim the Streets' and 'Critical Mass' have had more limited impacts.

Third, just as critiques of automobility have revealed its social inequities, so the social justice implications of strategies to reduce automobility must also be considered, especially in societies where it is deeply embedded. For example, in the absence of policies to mitigate their impact, higher gasoline taxes intended to reduce driving and internalize social and environmental costs hurt lower income more than higher income drivers (as the former spend a higher proportion of their incomes on transportation); the same is true of congestion charges and road tolls; and strategies to encourage a walking-city drive up real estate values in central cities, making housing there unaffordable for low-income people and increasing their dependence on cars at the same time as more affluent people are able to walk or cycle to work.

Finally, do alternatives to automobility involve 'technical fixes' (redesigning the material world) or 'social fixes' (changing social life and relations of power)? As several chapters suggest, strategies for change are likely a combination of both, even if they emphasize one over the other. For example, Henderson's discussion of contesting automobility through denser urban spaces and cultural change implies 'social fixes'. Social fixes would consist of convincing people to drive less, buy smaller, lighter, more fuel-efficient vehicles and live in denser urban spaces to enable other forms of mobility. Or, they would involve states making changes in the physical context of people's daily lives: stopping or even reversing urban sprawl, providing more and better public transit, and the like.

The concluding chapters in this volume focus specifically on the possibilities of societies shifting away from car dominance. Litman's analysis of mobility as a prestige good argues that social fixes require policy modifications of status incentives – in which the non-use of cars becomes 'cool' rather than a source of stigma, and public transit or self-propelled auto-mobility become associated with the good life. In suggesting alternatives such as the expansion of public transportation, Litman uses the rule of thumb that good public policy favours necessities over luxuries. Martin suggests that the shift away from auto dominance requires disrupting the strong connection between automobility, 'progress' and 'modernity' (at the societal as well as individual level). Even in China's relatively closed political system, dissent is growing against the regime's all-out push for motorization as a prime driver of economic development. Drawing on complexity theory, Dennis and Urry instead claim that public mobility, in which buses, trains and ships dominate, has been irreversibly lost. However, change has become so fluid that it is creating hybrid forms in which cars are becoming less privatized while public transportation models itself upon the flexibility and efficiency of car-ownership. More broadly, they contend that change can come about unpredictably and from unexpected sources. They do not entirely discount political action, but suggest that post-car mobilities will result from specific tipping points, moments of rapid change when a complex system switches.

The challenges facing those who would radically change automobility should not be minimized. But as this volume demonstrates, automobility is full of antagonisms, contradictions and complexities, and, thus, not inevitable. Alternatives are possible, but in sorting through pathways to change it is essential to understand how the automobile and its vast networks dominate mobility, why the car has wide appeal and what other forms of mobility will serve people in all walks of life, better.

Organization of the Book

As suggested by the title, the fundamental idea behind this book is that the car-system as the dominant mode of travel is problematic. The first section sets the stage by examining significant cultural meanings of automobility to help explain its dominance and troubles. The chapters range from a focus on automobile enthusiasts who run up against 'car troubles' in a specific location and historical period, to an analysis of how the representations of car ads draw on deeply-held cultural codes, to an exploration of how SUV ads have used design and photographic techniques to promote a phenomenal growth in sales, to an historical overview of the entrenched material culture of concrete in the formation of automobility. The second section considers the strategies and politics of regulating the dangers and risks of automobility, including the interventions of states, automakers and local communities. The third section highlights political contestations over automobility and raises the question of its inevitability. Chapters here range from an examination of local politics in Altanta, to national development strategies in Chile, and a theoretical analysis of the role of state and corporate power in making automobile travel 'compulsory'. The final section revisits issues raised in previous sections by examining obstacles to change based in status dynamics, the worldwide expansion of auto hegemony, and the 'locking-in' of automobility; it considers alternatives that go beyond the car.

References

Alvord, Katie. 2000. *Divorce your car! Ending the love affair with the automobile.* Gabriola Island: New Society Publishers.

Augé, Marc. [1992] 1995. *Non-places: Introduction to an anthropology of supermodernity,* John Howe, trans. London: Verso.

———. [1986] 2002. *In the metro.* Translated and with an introduction and afterword by Tom Conley. Minneapolis: University of Minnesota Press.

Beck, Ulrich. 1992. *Risk society: Towards a new modernity,* Mark Ritter, trans. London: Sage.

Beckmann, Jörg. 2004. Mobility and safety. *Theory, Culture and Society* 21(4/5): 81–100.

Böhm, Steffen, Jones, Campbell, Land, Chris and Paterson, Matthew, eds. 2006a. *Against automobility.* Cambridge: Blackwell Publishing.

Böhm, Steffen, Jones, Campbell, Land, Chris and Paterson, Matthew. 2006b. Introduction: Impossibilities of automobility. In *Against automobility,* Steffen Böhm, Campbell Jones, Chris Land and Matthew Paterson, eds, pp. 1–16. Cambridge: Blackwell Publishing.

Bradsher, Keith. 2002. *High and mighty: The dangerous rise of the SUV.* New York: Public Affairs.

Dunn, James A. 1998. *Driving forces: The automobile, its enemies, and the politics of mobility.* Washington: Brookings Institution Press.

Durkheim, Emile. [1951] 1966. *Suicide: A Study in Sociology*, John A. Spaulding and George Simpson, trans. Edited and with an introduction by George Simpson. New York: Free Press.

Edensor, Tim. 2004. Automobility and national identity: Representation, geography and driving practice. *Theory, Culture and Society* 21(4/5): 101–20.

English, Bob. 2008. Big hybrid still packs a performance punch. *Globe and Mail* 21 February. http://www.theglobeandmail.com/servlet/story/LAC.20080221. WHLEXUSHYBRID21/TPStory (accessed 22 February 2008).

Featherstone, Mike, Thrift, Nigel and Urry, John, eds. 2005. *Automobilities.* Thousand Oaks, CA: Sage.

Flink, James J. 1988. *The automobile age.* Cambridge: MIT Press.

Frank, Lawrence D., Andresen, Martin A. and Schmid, Thomas L. 2004. Obesity relationships with community design, physical activity, and time spent in cars. *American Journal of Preventive Medicine* 27(2): 87–96.

Freund, Peter and Martin, George. 1993. *The ecology of the automobile*. Montreal: Black Rose Books.

Gartman, David. 2004. The cultural logics of the car. *Theory, Culture and Society* 21(4/5): 169–95.

Garvey, Pauline. 2001. Driving, drinking and daring in Norway. In *Car cultures*, Daniel Miller, ed., pp. 133–52. Oxford: Berg.

Gitlin, Todd. 1986. We build excitement. In *Watching television*, Todd Gitlin, ed., pp. 136–61. New York: Pantheon.

Hannam, Kevin, Sheller, Mimi and Urry, John. 2006. Editorial: Mobilities, immobilities and moorings. *Mobilities* 1(1): 1–22.

Insurance Institute for Highway Safety (IIHS). 2005. Special issue: Vehicle incompatibility in crashes. *Status Report* 40(6) 28 April.

Jacobs, Jane. 1961. *The death and life of great American cities*. New York: Random House.

Jain, Sarah S. 2005. Violent submission: Gendered automobility. *Cultural Critique* 61: 186–214.

Jain, Sarah S. Locklann. 2004. 'Dangerous instrumentality': The bystander as subject in automobility. *Cultural Anthropology* 19(1): 61–94.

Keats, John. 1958. *The insolent chariots*. Philadelphia: Lippincott.

Kline, Ronald and Pinch, Trevor. 1996. Users as agents of technological change: The social construction of the automobile in the rural United States. *Technology and Culture* 37: 763–95.

Koshar, Rudy. 2004. Cars and nations: Anglo-German perspectives on automobility between the world wars. *Theory, Culture and Society* 21(4/5): 121–44.

Lomasky, Loren E. 1997. Autonomy and automobility. *Independent Review* 2(1): 5–28.

Lupton, Deborah. 1999. Monsters in metal cocoons: 'Road rage' and cyborg bodies. *Body and Society* 5: 57–72.

Lynd, Robert S. and Lynd, Helen Merrell. 1929. *Middletown: A study in American culture*. New York: Harcourt, Brace and Company.

———. 1937. *Middletown in transition: A study in cultural conflicts*. New York: Harcourt, Brace and Company.

McCarthy, John D. 1994. Activists, authorities, and media framing of drunk driving. In *New social movements: From ideology to identity*, Enrique Larana, Hank Johnston, and Joseph R. Gusfield, eds, pp. 133–67. Philadelphia: Temple University Press.

McLaren, Arlene Tigar. 2007. Automobilization and traffic safety. In *Reading sociology: Canadian perspectives*, Lorne Tepperman and Harley Dickinson, eds, pp. 290–93. Don Mills, ON: Oxford University Press.

McShane, Clay. 1995. *Down the asphalt path: The automobile and the American city*. New York: Columbia University Press.

Miller, Daniel, ed. 2001a. *Car cultures*. Oxford: Berg.

Miller, Daniel. 2001b. Driven societies. In *Car cultures*, Daniel Miller, ed., pp. 1–33, Oxford: Berg.

Mills, C.W. 1959. *The sociological imagination.* New York: Oxford University Press.

Mumford, Lewis. 1964. The highway and the city. In *The highway and the city*, Lewis Mumford, pp. 176–89. London: Secker and Warburg.

Nader, Ralph. 1965. *Unsafe at any speed; The designed-in dangers of the American automobile.* New York: Grossman.

Newman, Peter and Kenworthy, Jeffrey. 1999. *Sustainability and cities: Overcoming automobile dependence.* Washington, DC: Island Press.

Norton, Peter D. 2008. *Fighting traffic: The dawn of the motor age in the American city.* Cambridge: MIT Press.

Packer, Jeremy. 2008. *Mobility without mayhem: Safety, cars, and citizenship.* Durham: Duke University Press.

Paterson, Matthew. 2007. *Automobile politics: Ecology and cultural political economy.* Cambridge: Cambridge University Press.

Peden, Margie, Scurfield, Richard, Sleet, David, Mohan, Dinesh, Hyder, Adnan A., Jarawan, Eva and Mathers, Colin, eds. 2004. *World report on road traffic injury prevention.* Geneva: World Health Organization.

Reinarman, Craig. 1988. The social construction of an alcohol problem: The case of Mothers Against Drunk Drivers and social control in the 1980s. *Theory and Society* 17: 91–120.

Roy, Francine. 2007. From roads to rinks: Government spending on infrastructure in Canada, 1961 to 2005. *Canadian Economic Observer* 20(9): 3.1–3.22.

Sachs, Wolfgang. [1984] 1992. *For love of the automobile: Looking back into the history of our desires,* Don Reneau, trans. Berkeley: University of California Press.

Sahlins, Marshall. 1976. *Culture and practical reason.* Chicago: University of Chicago Press.

Sheller, Mimi and Urry, John. 2000. The city and the car. *International Journal of Urban and Regional Research* 24: 737–57.

——. 2003. Mobile transformations of 'public' and 'private' life. *Theory, Culture and Society* 20: 107–25.

Tarrow, Sidney. 1998. *Power in movement: Social movements and contentious politics, Second edition.* New York: Cambridge University Press.

Urry, John. 2004. The 'system' of automobility. *Theory, Culture and Society* 21(4/5): 25–39.

——. 2006. Inhabiting the car. In *Against automobility*, Steffen Böhm, Campbell Jones, Chris Land and Matthew Paterson, eds, pp. 17–31. Cambridge: Blackwell Publishing.

Vallières, Lucie. 2006. Disciplining pedestrians? A critical analysis of traffic safety discourses. Unpublished M.A. Thesis. Vancouver: Department of Sociology and Anthropology, Simon Fraser University.

Waldman, Amy. 2005. India accelerating. *New York Times* 4–7 December. http://www.nytimes.com/ (accessed 8 December 2005).

Williamson, Judith. 1978. *Decoding advertisements: Ideology and meaning in advertising.* London: Boyars.

PART 1
Cultures of Automobility

Chapter 1

T-Bucket Terrors to Respectable Rebels: Hot Rodders and Drag Racers in Vancouver BC, 1948–1965

Catharine Genovese

The experience of auto-mobility has a vast range of manifestations in diverse cultural practices. Drag racing is only one cultural expression that emerged with the development of the automobile. But it illustrates in a particularly dramatic form the process by which communities in specific contexts negotiated and constructed both the car's seductive appeal and its troubles. While the car provided opportunities for power, speed and excitement for young, white working-class men, their creative modification of cars troubled the wider community, symbolizing rebellion and outlaw behaviour on the streets. This case study shows how a drag racing community struggled over meaning, rules and regulations in the face of public disapproval, and how it sought resolution between paradoxes of automobility (e.g., danger and safety, the outlaw and the respectable, freedom and regulation, the magical and the mundane) by policing its own boundaries and seeking to create a respectable hot rodder identity. In their active negotiations with the wider community and transformation of their own identities, this relatively powerless group exercised agency through auto-mobility practices in multiple social sites and contributed to cultural meanings and technological innovations during a time of rapid expansion of automobility.

The movie *Rebel without a Cause* (1955) established drag racing as a symbol of youthful rebellion in the United States, and young working class males in Vancouver, BC, were a part of this transnational movement. The impact of the automobile on North American society has been integral in the development of modernity and its urban and suburban landscapes. With its celebration of speed and excitement, drag racing contributed to the growing interest in the automobile. The defining features of drag racing in its early stages are captured by the National Hot Rod Association: two high-stepping coupes blast off the starting line in the 1950s in a quarter-mile celebration of acceleration (NHRA 2001). Drag racing is a sport that grew out of hot rodding and the desire of hot rodders to compete against like-minded performance enthusiasts in a professional forum. Hot rodders modified whatever vehicle they could get their hands on, stripped it down for speed, performance and a particular look and competed in events with or without official sanction.

The popular view portrays the establishment of drag racing and hot rodding in North America after World War II as a natural, linear progression resulting from the coincidence of automobiles, ingenuity and opportunity. "Illegal street racing had been around as long as the automobile, but it took off after World War II, when GIs with enhanced mechanical skills and a love of speed and danger returned to civilian life" (NHRA 2001, 19). Such depictions view drag racing as an inspiring example of American know-how, creativity, and the utility of capitalism to satisfy the emotional needs of the individual. But the development of drag racing cannot be explained simply by emotional drives or commodity consumption, and it has not been a straightforward march towards ever-increasing speed records. In Vancouver it was also a story of transforming rebels into respectable citizens. The British Columbia Custom Car Association (BCCA) was arguably the first club in Canada to interpret and develop the guidelines provided in *Hot Rod* magazine to begin their own club and eventually become the state-sanctioned owners of Mission Raceway Park in Mission, BC (Mission Raceway Park 2005).

The period 1948–65 merits scrutiny as the period when drag racers and hot rodders across North America developed an internal bureaucracy to counter negative portrayals in the media and public derision, and to deflect unwanted attention from government and law enforcement officials. In doing so, the movement changed from rebel protest to respectable accommodation. The outcome of this shift would be social tolerance and opportunities for some members to capitalize economically on the sport. Unlike Derek Simons (this volume), I do not examine the historical and ontological origins of speed. Instead, based on documentary evidence and interviews with 12 Vancouver-area hot rodders and related stakeholders, I provide a case study of how a group's interest in speed was intertwined with a social and legal context. As the group negotiated this context, they defined a legitimated, respectable and authentic identity – the 'true hot rodder' – which included both their interest in speed, and their desire to perfect technical performance and driver skills.

The movement began after World War II in Southern California with the publication of *Hot Rod* magazine and the formation of the sport's primary governing body, the National Hot Rod Association (NHRA), and extended quickly throughout North America. The BCCA in Vancouver, BC, was one of the groups receiving, interpreting and reinterpreting the NHRA's edicts through *Hot Rod* magazine. *Hot Rod* magazine was where the enthusiast community, street racers and the public learned what it was to be a true (read safe) hot rodder. The hot rod movement sought to distinguish itself from those involved in the 'traffic light grand prix': mainly youths, who raced main thoroughfares, injuring and killing themselves and innocent bystanders (Smith 2006). Harry DeSilva wrote in his 1942 book, *Why We Have Automobile Accidents* that,

> Present day youth has developed nocturnal habits which from a safety standpoint are highly undesirable ... Nowadays a young man thinks he must borrow the family car, collect a group of friends and rush all around the country looking

for entertainment. Often encouraged by alcoholic refreshments he is lured into demonstrating to his companions how fast 'the old bus' will go, with disastrous results. (DeSilva 1942, 205)

Since the general public did not recognize the distinction between drag racing and street racing, its negative glare was cast towards both.

Thus far the limited historiography available on the subject of hot rodding has focused on the structure of its governing bodies, technological aspects of the sport activity, and the significance of the message being distributed. In his analysis of the culture and technology of drag racing from 1950 to 2000, Robert Post (2001) examines how the technological quest for speed drives the sport of drag racing. Bert Moorhouse (1991, 39) asserts that "[t]o counter bad publicity and the effects this might have on the sport and the burgeoning economic interests" associated with it, the emerging governing apparatus "aimed at incorporating street racing into the serious activity" of drag racing. My research seeks to uncover the ways in which a Canadian club understood, construed and utilized the guidelines from *Hot Rod* magazine on club formation and how the club established drag racing and hot rodding in Vancouver as a legitimate sport and enthusiasm. I examine how the information, received from the magazine, was reconstructed to enable members of one hot rodding club in Vancouver to become respectable rebels who could pursue their sport as they desired, and who would become capable of reaping the economic benefits of their enthusiasm.

Although many popular historical books with exciting pictures focus on hot rodding and drag racing, critical historical work on the rise of drag racing has been less forthcoming. Traditionally, historians assumed that 'popular culture' was less significant socially than 'high culture'. They made inherent class distinctions between high culture as art and popular culture as mass-produced commodity that gauges success by commercial accomplishment (Storey 1997). In the 1990s cultural historians became critical of such a distinction that implied people were "cultural dopes" lacking the ability to assess what is meaningful or important and "thus at the economic, cultural, and political mercy of the barons of the industry" (Fiske 1998, 504). John Fiske (1998, 505) argues persuasively against the image of the cultural dope, contending that "despite the homogenizing force of the dominant ideology," subordinate groups have utilized the diversity of capital to produce an exponential number of voices and this variety of voices allows these people to compare and decide on the capitalist forms to which they choose to relate. Historians have since stepped away from this homogenous view and do not see the social world as a dichotomy between the high and the low. They are more inclined to take Stuart Hall's (1998) position that emphasizes a dialectic between the two alternative poles of containment and resistance and the social space within which these two poles are constantly coming together and being superimposed upon each other while society negotiates from within to determine quotidian practices. Additionally, far from seeing popular culture as only the "expressive culture" of the masses, recent works are beginning to relate popular culture to power and "recognizing that popular culture cannot be defined in terms of its intrinsic properties but must be conceived in relation to the political forces and cultures that engage it" (Joseph and Nugent 1994, 15).

Historical and ethnographic studies have begun to examine the car within the cultural and political context of specific communities (see, for example, Miller 2001). Given the epoch-making impact of the automobile, Gilroy (2001) notes, it is imperative to examine how black communities respond and participate in its consumption. In particular, he explores how cars are uniquely linked in the lives of black communities to their "broadest political and economic hopes" (82). Other studies, highlighting the interaction between the social and technological, examine how specific groups are active participants in the social construction of the automobile. Kline and Pinch (1996) show that during the early decades of the twentieth century farm people in the rural United States used and modified the car in ways not anticipated by manufacturers. In particular, the authors argue that because competence in operating and repairing machinery was central to their masculinity, farm men opened up the 'black box' of the car, reinterpreting its function and using it for varied purposes such as grinding grain. The authors adopt the concept of 'interpretative flexibility' to reflect the process in which specific social groups change technologies by investing them with new uses and meanings. By the early 1950s, the authors contend, such flexibility had disappeared as farm people used cars as intended by manufacturers. This chapter illustrates that interpretative flexibility persisted in the 1950s as urban and suburban working-class men turned the automobile to their own purposes of self-expression, in the form of hot rodding, and consequently contributed to changes in its technology. In this process, they also developed and transformed their identities.

The enthusiasm of these young men for drag racing cannot be understood in traditional narratives that see popular cultural signifiers as essentialized customs applicable to the entire community. The North American drag racing community's identity was not monolithic and the quotidian practices of the group continued to redefine the values, language, and interpreted traditions of the community. Additionally, new members coming into the group contested existing practices and power structures to produce new cultural expressions within the racing community. These new expressions presented themselves to the public at the strip and on the street. The constant evolution and redefinition of the expressions of culture and community through practice and language resulted in the transformation of members of Vancouver's BCCA from problematic teenagers into respectable citizens, professional sportsmen, and in some cases, business managers (Carey 1996).

Drag racing's early development as a popular cultural form entails one of the central, defining production commodities of modernity and reveals trends and social attitudes that continue to shape reaction to the automobile to the present. This history of drag racing begins with the mass production of the car, as it was in part the impetus for the hot rod. Henry Ford began his groundbreaking assembly line production of the Model T before World War I with the professed intention of making it available to the common man. But as Warren Susman (1984) points out, Ford did not realize that the common man did not want to feel common. In his drive to mass produce and mass market a commodity that utilized the assembly line, slick new advertising techniques and a state-sponsored system of highways and roads, Ford failed to perceive that, "mechanical perfection, although desirable, was not enough" (Susman 1984, 140). Ford's assembly line did, however, lay the foundation for certain people – especially young white males – to purchase mass produced cars and use their interpretative flexibility for self expression.

Although 'hot rodding' or modifying cars began soon after Ford supplied them, the specific date of the transformation of drag racing from hobby to sport is difficult to ascertain. As Wally Parks, the most famous editor of *Hot Rod* magazine said, "street racing has probably been around since the first two owners of horseless carriages lined up to see whose mount was fastest" (NHRA 2001, 13). By 1948, enthusiasts across North America were modifying their cars for looks and more importantly, for speed, following the trends of the emerging California car culture. The Southern California Timing Association (SCTA), established in 1937, began organizing drag racing on the flat, dry salt lakes that were the safest, most reliable areas to test their vehicles (Post 2001). The hot rod apparatus that Moorhouse (1991) describes consisted of the SCTA membership, who after WWII would leave the dry lakes to find decommissioned airstrips, form the NHRA and become the writers and editors of *Hot Rod* magazine. With these modifications, street racing intensified and public attention to it grew, as illustrated by a number of high profile articles in publications such as *Colliers*, *Life* and the *New York Times* (Moorhouse 1991). Widespread popular media coverage of hot rodding and street racing was testament to the number of racers, fans and public attention that the enthusiasm was attracting, and a demonstration that drag racing was not just a fad.

Initially, Vancouver hot rodders represented trouble to law enforcement and society, as uncontrolled youth on the streets. The automobile enabled young people to cruise around in public without adult supervision; it was easily modified for style to project personal taste, and it could be 'hopped-up' for speed. As Conley argues (this volume), the car has often represented freedom, in this case encompassing all that was important to postwar society. Like other cities across North America, Vancouver had its share of street racing. Local police and the RCMP were constantly breaking up clandestine street races across the Lower Mainland in Tsawwassen, up and down Burnaby Mountain, at Marshland Avenue in Burnaby, at Port Kells in Surrey, and on the Lougheed Highway near Mission. Some of these locations were hardly developed at the time; few people were around, and drivers had many opportunities to race illegally. These races occurred in addition to the daily incidents of racing that the police claimed they were encountering on major thoroughfares. The newspapers are dotted with reports such as an 1 October 1952 article in *The Vancouver Sun*, where the BCCA urges parents to curb young drivers, and the rodders themselves have told of crashes, rollovers, cars in ditches, and drivers hurt or killed (Warren 2005; Jeboult 2004). Cities across North America were experiencing a moral panic over juvenile delinquency and the images of hot rodders as youth alienated from the larger community found in cautionary films such as *Hot Rod Rumble* (1957), *The Ghost of Dragstrip Hollow* (1959) and *Dragstrip Girl* (1957) (Moorhouse 1991). A Martens' 1950 play, *Drag Race: A Teenage Play in One Act* depicted drag racers as unrepentant teenagers who only became penitent when they realized that the young man they were racing against had been killed.

Most hot rodding and drag racing enthusiasts realized that control, safety and accommodation would be the only way to counter the negative publicity and state oppression of their sport. To give their own members a forum for discussion, the racers and rodders in California began publishing a magazine, *Hot Rod*, which would become the voice of the drag racing enthusiast community. *Hot Rod's* readership, which began at 5,000 monthly in 1948, grew rapidly to 200,000 monthly across the United States and Canada by the end of 1950 (Moorhouse 1991, 41). The California racers who wrote and edited *Hot Rod* clearly believed that they had become respectable, and their aim was to promote hot rodding and racing as a reputable enthusiasm and professional sport. *Hot Rod* magazine's role in the promotion of the enthusiasm and its inherent safety would be clarified when in 1949, the *New York Times* ran a statement from Thomas W. Ryan, the Director of the New York Division of Safety who said,

> Possession of the 'hot rod' car is presumptive evidence of an intent to speed. Speed is Public Enemy No. 1 of the highways. It is obvious that a driver of a 'hot rod' has an irresistible temptation to 'step on it' and accordingly operate the vehicle in a reckless manner endangering human life. It also shows a deliberate and premeditated idea to violate the law. These vehicles are largely improvised by home mechanics and are capable of high speed and dangerous maneuverability. They have therefore become a serious menace to the safe movement of traffic. The operators of these cars are confused into believing that driving is a competitive sport. They have a feeling of superiority in recklessly darting in and out of traffic in their attempt to outspeed other cars on the road. (Moorhouse [1950] 1991, 354)

In reply to these dramatized allegations voiced across North America, *Hot Rod* magazine replied in the next issue that,

> A real hot rod is a car that is lending itself to experimental development for the betterment of safety, operation, and performance, not merely a stripped down or a highly decorated car of any make, type of description, or one driven by a teenager. As to the menace or nuisance element, very few hot rod enthusiasts want to risk their specialized equipment for use as battering rams. The fact their cars are built so that they attract attention become an automatic psychological brake which governs their driving activities. (Moorhouse [1950] 1991, 354)

Hot Rod was attempting to define the emerging sport, to identify what it meant to be a hot rodder in professionalized and technical terms. In addition, the commentary took pains to distance the hot rodder's identity from the publicly-maligned teenager and to encourage the illegal racers to conform to the public-friendly regulations that would allow drag racers to be accepted by a mass audience. As Kline and Pinch suggest, in specific contexts, users socially construct technologies. In this case, by stressing the experimental development of the car, the magazine not only provided a flexible interpretation of the automobile vis-à-vis manufacturers. It also indicated that different social groups within the hot rodding community held distinct views about what the cars mean. Interpretative flexibility was also practiced at the local level of drag racing.

Intent on their discussions of regularization and representation, the drag racers and rodders in California decided that a national organization for racers should be established. In 1951 they formed the National Hot Rod Association (NHRA) to increase safety, codify rules, and to attempt to integrate street racers into a more professional arena of racing on the strip. By professional, the NHRA intended to represent all conforming enthusiasts as law-abiding and respectable. Their desire for accommodation into mainstream society was driven by the aspiration for state sanction rather than surveillance, and by the ambition of some members to be the arbiters of the internal bureaucracy. "Regarding the institutional structure of drag racing, power brokers like Wally Parks saw an opportunity to establish their authority to make rules and confer or withhold sanction according to their own precepts" (Post 2001, 11).

The culture produced for the community was strategically disseminated for the purposes of quieting public complaint, creating an internally unified identity and enabling those qualified enthusiasts to capitalize on the emerging 'speed' market that required specialized mechanical and managerial skills. Street racers who would not conform to NHRA structures could then be labelled outlaws who compromised public safety. If the public maligned the street rodder as unsafe and irresponsible, NHRA officials could shake their heads and agree. By withdrawing their sanction they could doubly isolate 'the outlaw', and protect their pursuit as a sport, rendering it safe in a vehicular and social sense, and thus have state protection for the organization and its leaders. This position vis-à-vis the state and law enforcement protected the community and more importantly, the NHRA from outside interference. Such a strategic move demonstrates how this popular culture of hot rodding developed within specific legal, political and social contexts. As long as the NHRA put up the public image of safety and continued to make money to support itself, it was free to develop the sport of drag racing in the direction of increasing technological innovation, racing for speed and records, and producing more publicly accessible parts and specialty skills that could economically enrich its members. Hot rodders and street racers who wished to continue pursuing unsafe and illegal racing were not only subject to state oppression and differentiation from the government-sanctioned racers; they also lost the opportunity to capitalize on rapidly expanding economic opportunities.

Other strategies to code rules included developing physical space with its own social meaning. The racetrack, in particular, became the most important signifier of respectability. To convince young hot rodders and street racers that it was more credible within the community to test their cars on the strip, voices in *Hot Rod* magazine continued to educate enthusiasts not only about the hundreds of technical issues but also about a variety of social issues such as: how to form a club, ways to fundraise, how to obtain local areas for racing such as airstrips, how to work with local police, and most importantly, how to convince the general public that hot rodders were not hooligans. "The magazine suggested that the timing associations (clubs) should makes themselves more attractive to prospective members and that the law courts should take a much tougher attitude" (Moorhouse 1991, 41[HRM 9 September 1956, 5]).

Rodders and racers in Vancouver had a blueprint in the form of *Hot Rod* magazine, available in the Lower Mainland since its initial issues in 1948. Using the models provided in *Hot Rod* magazine, the British Columbia Custom Car Association (BCCA) was formed with a handbook that stressed respectability and the desire to avoid negative public labelling that it believed would hurt hot rodding. On 2 January 1952 the club held its first official meeting and elected a President, Vice President and Secretary/Treasurer. They immediately began publishing a newsletter called 'Clutch Chatter' and on 9 December 1952, the club became registered as a non-profit association and incorporated under the Society Act of British Columbia, which was a significant indicator of state sanction and propriety (Mission Raceway Park 2005). The fee to join was $5 and yearly dues were $3 (Farmer 2004; Jeboult 2004). The average Canadian hourly wage in the construction and manufacturing industries at the time was $1.48 (Statistics Canada 2006) and the leadership of the BCCA wanted their young, working class peers to be able to afford to join. "Many young accident repeaters come from the lower economic class whose families cannot afford a car, and who, consequently, drive borrowed autos, commercial vehicles on off-time, or jalopies owned jointly with several other penniless young fellows" (DeSilva 1942, 206). Much like *Hot Rod* magazine itself, these Vancouver rodders were concerned with the troubling connotations that surrounded the words 'hot rod' and so the BC Custom Car Association was selected as the name that could encompass all types of hot rods succinctly (Jeboult 2004). This was necessary because the public viewed hot rodders as petty criminals racing about on the streets. Tired of traffic tickets and infamy, the BCCA aimed to change that perception to gain respectability and obtain a place to race. Thus, at a local level, the hot rodding community, which contained a variety of voices in relation to the bourgeois forms of the automobile, negotiated both internally and externally. As such, it illustrates the theoretical argument that popular culture is not simply an intrinsic property of the 'lower' classes, but develops in relation to political and cultural forces with which it engages (Joseph and Nugent 1994). Chief among these were the police, the public, and later, politicians.

Taking their cues from *Hot Rod* magazine guidelines, the members of the BCCA set out to get local law enforcement on their side. They approached Chief

Mulligan of the Vancouver Police directly, and he assigned two young officers to become part of the Vancouver Traffic and Safety Council that would meet with the rodders and assist them with obtaining a place to race (Jeboult 2004). They also set out to win over the public, as they were keenly aware that public disapproval would force the police and government to act against them. Club members were encouraged to help motorists in distress and were given cards to hand to the relieved commuters that read:

> You have been assisted by a member of The British Columbia Custom Car Association. A 'Hot Rod' organization formed by a group of responsible auto enthusiasts, dedicated to promote interest in the sport, wherever it may be found, and which some day hopes to unveil to the public the true meaning of the word 'Hot Rod'.

Coming straight from the pages of *Hot Rod* magazine, such tactics illustrated the ways in which the BCCA was striving to promote and legitimize their sport. Positive reaction to the club's formation was immediate, and from a small group of enthusiasts in 1952, the BCCA's membership grew to over 300 by 1954, including more than 200 rod and custom cars as well as 11 drag race vehicles (Mission Raceway Park 2005).

Legitimation meant playing by the rules. If hot rodders wanted to become members of the BCCA, they had to pay their fees, abide by very stringent safety rules, wear a helmet, and have their cars 'teched' by a licensed mechanic prior to the race. If a car did not meet the required standards or did not contain the safety equipment required for its speed class, the racer would not be allowed to compete. Drag racers also had to agree to race only on specific, sanctioned days, and make only a limited number of runs (Jeboult 2004; Farmer 2004). These rules and regulations were crucial if the BCCA were to win state and public acceptance. Controlling the outlaw street racers who threatened public safety and social order was a critical strategy and the price the association was willing to pay. Street racers, on the other hand, argued that the regulations detracted from the spontaneous emotional gratification that came from illegal racing (Jeboult 2004; Smith 2006). Such ongoing struggles of legitimacy included renegotiating meanings of masculinity. The BCCA, like the American NHRA, sought to substitute a different form of emotional payback, through the identity of the 'real man' who was mechanically resourceful and responsible. For example, as a teenager Keith Warren (2005) was sent by his parents to private school to get him away from cars and the buddies he made in the hot rodding community. Two years later, his approach changed as he joined the BCCA, began working for the Royal Bank, and became a member of Jack Williams' famed Syndicate Scuderia pit crew. The 'true', safe hot rodder, as defined in the rulebook of the BCCA and the pages of *Hot Rod* magazine, would not risk his life or his car on the streets. Racing and the associated industry would flourish only if it condoned 'legitimate' racers as safe, responsible, skilled professionals out to fulfill their dreams within the dominant culture, not in opposition to it. Thus, hot rodding, one aspect of North American automobilization, which was almost purely 'magical' had to incorporate the 'mundane' in the words of Jim Conley (this volume).

Authorized by provincial laws and statutes, the BCCA had the power to define who were legitimate hot rodders, determined by their club membership and their adherence to a created set of traditions, language and practices that defined the community. Where winning the street race was formerly the sole determinant of status, success was now gauged by persistence, safety and constant innovation. Thus, the exposure of the objectives of subcultures to the society at large sheds light on the accepted stereotypes held by those outside the community. This examination of the everyday experience of drag racing suggests that through the language of professionalism, the evolving traditions of membership, and the symbols of drag racing, the BCCA used the guidelines put forth in *Hot Rod* magazine to gain respectability and autonomy.

Drag race enthusiasts did not, however, pursue regulation and respectability merely to protect the sport. The interpretative flexibility of the automobile (as in the case of the farmers in Kline and Pinch's 1996 study), opened up a market for both auto manufacturers and makers of parts and accessories. Drawing on its extensive economic connections in the automobility system, drag racing developed a distinct economic motivation. In the United States in 1950 alone, it was estimated that the bolt-on parts industry, the manufacturers of equipment, their distributors, speed shops, professional mechanics, and other retail interests grossed eight million dollars (Balsley 1950). As drag racing gained official permission, its commercial potential increased; money was to be made in owning strips, managing and promoting races, and sponsorships (Post 2001). Racing enthusiasts who had the driving, mechanical, and managerial skills could capitalize on the demands made by the rest of the public to build cars that could be driven down to the strips and raced. There was profit to be made from the success of drag racing. Such economic motives led to further self-regulation. Just as *Hot Rod* magazine and the NHRA were creating a sport and a space for a 'legitimate hobbyist', the BCCA was also establishing the corresponding economic framework which would make some of its members respected businessmen, leaders in their technological fields, spokespersons representing the sport to the government, and ultimately, gatekeepers who determined who was in and who was out. The members of the BCCA realized soon after it began that having a track was paramount to their success as racers, and for some, as small capitalists.

The club knew early on that having their own space would legitimize their sport, and gain them public respect. With assistance from the articles in *Hot Rod* magazine and guidance from their police officers, the BCCA began fundraising for a track in earnest and brought in money by showing cars to the public. In late 1952 the club hosted what it claims to be Canada's first Rod and Custom Show at the Kerrisdale Arena, then one of Vancouver's premier venues and a step up from impromptu car shows at the local drive-in. The show eventually became known as the Pacific International Motorama and ran annually until 1973 at the Pacific

National Exhibition ground as part of Vancouver's mainstream cultural mix. The club also held 'Reliability Runs', in which the drivers had to complete timed routes with checkpoints where spectators could view 150 to 200 cars. In those days not many cars had fancy paint jobs, with most of them in primer and without fenders. This is the era when the adage, 'run what you brung' originated, as more often than not cars were cobbled together in shops where the club members would pool their skills and tools to get a car on the road or on the track. How the car looked was not as important as reliability and speed (Jeboult 2004; Smith 2006; Warren 2005).

In 1952, BCCA began hosting the first organized drag races in Canada at the Abbotsford Airport. Spectators were kept on one side of the airstrip and the pits were located on the other. Races were started by a club member flourishing a green flag and times were clocked by members' wives and girlfriends stationed in the club bus at the end of the track with stopwatches (Jeboult 2004). To this day, the constitution of the BCCA states that a member is a man of at least 16 years of age who holds a valid driver's license. Women were allowed to race but they have never been members of the BCCA (Warren 2005). The girlfriends and wives of the members eventually became the BCCA Ladies Auxiliary and their duties were circumscribed to fundraising, helping out at the track during races, and cleaning the airstrip once the racing was finished. During this time, the BCCA was getting increasing attention in the United States from various rod and custom magazines with articles such as Tommy Amer's in *Hot Rod* (October 1954), and *Hop Up's* "Canadian News" (December 1952), "North of the Border" (February 1953) and "Canadian Capers" (January 1953).

Racing at Abbotsford lasted until 1957 when the Royal Canadian Air Force took over the facility and the BCCA began to move to purchase land to operate a track. During the 1950s the club banked every dollar made at the Abbotsford drag races and other events for a permanent drag strip. In 1959, the club located an appropriate piece of property southwest of the City of Mission along the banks of the Fraser River. The *Fraser Valley Record* (30 March 1960) reported that clearing, surveying and grading were completed in 1960 but the period of good luck for the club was brief. For the next three years the BCCA would have to spend its hard earned 'track fund' money on property negotiations, legal surveys, and other procedures such as trying to convince the Councillors of Mission that a track would be both suitable and profitable. The BCCA's major emphasis had shifted from racing and rodding to legal and political battles, illustrating again, how specific types of cars and drivers troubled wider communities and required intense political negotiations. The town council of Mission was loath to the idea, believing the track would be noisy and that it would bring an unruly element to the streets of Mission. It took years of meetings, and even one alderman visiting the Arlington raceway in Washington State before the Mission Council would finally concede.

The BCCA's money was running out in 1963 when the BC Minister of Highways, the Honourable 'Flying' Phil Gaglardi lent his assistance by providing engineer's estimates, preparing the site for paving and finishing the 3550 foot

long and 60 foot wide track using his own crew (*Fraser Valley Record* 1965).
Mr Gaglardi got his nickname for using a government Lear jet to fly family and
friends around BC, but he was also a staunch supporter of motorsport venues.
Honourable 'Flying Phil' Gaglardi was also known for racing about in his 1958
Impala, 'testing the new highways' of the province for safety, he claimed, and
receiving frequent speeding tickets as a consequence (Warren 2005). In light of
the boom in construction and infrastructure around North America at the time, the
BCCA was fortunate to have a high-ranking cabinet minister supporting its efforts.
After 14 years of work, and at a cost of $50,000 according to club president Larry
Braine, the BCCA club finally had its own space (*Fraser Valley Record* 1965). The
first drag races were run at Mission Raceway on 26 August 1965. Although the
facility initially lacked bleachers, guard rails and fencing, and operated with a basic
single lane timer, the BCCA considered it to be a great success. Between 7,000 and
10,000 spectators, including the enthusiastic Phil Gaglardi crowded the track. The
BCCA and other track sponsors had only expected 2,000, so the streets of Mission
were packed with cars, and people sat up the hill of Lougheed Highway, and on
their roofs to get a good vantage point. Entrance fees for spectators were $2, and
an extra $1 for a pit pass to get a closer look at the cars (Farmer 2004). On opening
day, 130 cars, including many from Alberta and Washington, entered in the drags,
and top speeds hit 170 mph. The track was primarily run by BCCA volunteers,
who in 1965 received $10–15 per day as a gas allowance for their commute from
Vancouver to Mission and as a small payment for their efforts. In fulfillment of the
NHRA and its own safety requirements, the association also had to pay $25 a day
for the services of a 1960 International Ambulance and its attendants to sit at the
track in case of emergency (Farmer 2004).

Once the track was established, the BCCA sought to become sanctioned by
the NHRA so the track times and records could be made 'official' and compared
with other tracks in North America. Sanctioning required complying with the rules
and regulations set out by the NHRA. The club installed guardrails, purchased
Chrondek Timers to replace the Ladies Auxiliary's stopwatches, and built fencing,
bleachers, washrooms, concessions, and a proper pit area. A Christmas tree (the
vertical strip of lights to start races) replaced the flagman and in 1966, Mission
Raceway joined NHRA Division Six.

Because it was built below sea level, allowing cars to run faster, Mission
Raceway soon became known internationally for its outstanding traction (Mission
Raceway Park 2005). In 1967 Mission hosted its first NHRA World Championship
Series race, and in subsequent seasons more National records were set at the
Mission Raceway than any other track, which at one time held both the Top
Fuel and Funny Car National Records (Mission Raceway Park 2005). Becoming
sanctioned and respectable meant the members of the BCCA were operating
arguably one of the foremost tracks in North America, and certainly the fastest
track in NHRA Division Six, which was dubbed, 'Land of the Leaders'. The club
was able to fulfill successfully drag racing's objective of pursuing the limits of
speed under controlled conditions.

Regulated and controlled, drag racing steadily improved its safety record and pioneered advancements in the safety technology that contributed to the protection of the general public as car designers and engineers looked to the track for inspiration. Technology was improving quickly and the cars also became faster. For example, in the five years between 1950 and 1954, speeds of the fastest machines had improved by about 25 miles per hour (Post 2001). The more vehicles were modified for speed, the more specialized skills were required; most importantly, the faster and more innovative the car, the higher in class it became. The top speed and the lowest elapsed time down the track were the prime objectives, which were very different from street racing. Drag race enthusiasts considered a car modified for looks as a 'shot rod', more of a Sunday cruiser or 'grocery-getter' associated with suburban family life than a real 'hot rod' (Warren 2005).

Thus formation of the BCCA enabled Vancouver racers and rodders to develop their own space, attain a modicum of social esteem, and capitalize on the emerging economic potential surrounding their sport. Resistance had been transformed into accommodation and state sanction. By examining the development of the BCCA from rebellious youth to professional sportsmen and business managers in the early years, it can be seen that the transformation in the enthusiasm took place as much from internal struggles for definitions as from external pressures. The leadership of the BCCA needed to protect their interests, both financial and recreational, from negative public and state scrutiny. To do so, they created an identity for the Vancouver hot rodding community, interpreted from the pages of *Hot Rod* magazine that was safe, responsible, and separated from youths on the street. Contrasting the classifications in America's *Rebel Without a Cause*, the BCCA were featured in a National Film Board 1957 release entitled *Fun for All: Car Crazy*, in which the BCCA members were praised for being safe and responsible sportsmen. Between 1948 and 1965, the BCCA incorporated and accommodated society's values to protect its members, displaying motives such as protection of their community, the growth of an internationally state-sanctioned sport, and the capability to economically capitalize on the profitable speed industry. Through representation and adaptation, the rebels had become respectable. In this process, hot rodders reworked the cultural meaning of the car – its appeals and its troubles – in specific social contexts. They contributed to the experience of auto-mobility by transforming hot-rodding identities and redrawing the boundaries between legitimate and illegitimate driving, and to the system of automobility by creating new physical and social spaces and testing the technological limitations of speed. This analysis illustrates the intricate relationship between both cultural and material developments of the automobile during a period of interpretative flexibility.

References

Balsley, Gene. 1950. The hot rod culture. *American Quarterly* 2(4) winter: 353–58.

Carey, James W. 1996. Overcoming resistance to cultural studies. In *What is cultural studies?: A reader*, John Storey, ed., pp. 61–74. New York: St. Martin's Press Inc.

DeSilva Harry. 1942. *Why we have automobile accidents*. New York: John Wiley and Sons.

Farmer, Ron. 2004. Interview by author, 28 October, Langley, BC. Tape recording.

Fiske, John. 1998. The popular economy. In *Cultural theory and popular culture: A reader. Second edition,* John Storey, ed., pp. 495–512. Athens: University of Georgia Press.

Fraser Valley Record. 1960. $70,000 Drag-strip project for flats here. *Fraser Valley Record* 30 March.

—. 1965. Record crowds attend drag strip opening. *Fraser Valley Record* 1 September.

Gilroy, Paul. 2001. Driving while black. In *Car cultures,* Daniel Miller, ed., pp. 81–104. Oxford: Berg.

Hall, Stuart. 1998. Notes on deconstructing 'the popular'. In *Cultural theory and popular culture: A reader. Second edition,* John Storey, ed., pp. 442–53. Athens: University of Georgia Press.

Jeboult, Bunny. 2004. Interview by author, 15 October, Hatzic, BC. Tape recording.

Joseph, Gilbert M. and Nugent, Daniel, eds. 1994. *Everyday forms of state formation: Revolution and the negotiation of rule in modern Mexico.* Durham: Duke University Press.

Kline, Ronald and Pinch, Trevor. 1996. Users as agents of technological change: The social construction of the automobile in the rural United States. *Technology and Culture* 37: 763–79.

Miller, Daniel, ed. 2001. *Car cultures.* Oxford: Berg.

Mission Raceway Park. 2005. *Club history, Mission, BC.* http://www.missionraceway.com/mrphistory/history.htm (accessed 23 January 2005).

Moorhouse, Herbert F. [1950] 1991. *Driving ambitions: An analysis of the American hot rod enthusiasm*. New York: St. Martin's Press.

NHRA (National Hot Rod Association). 2001. *The fast lane: The history of NHRA drag racing.* New York: Reagan Books.

Post, Robert C. 2001. *High-performance: The culture and technology of drag racing, 1950–2000. Revised edition.* Baltimore: Johns Hopkins University Press.

Smith, Bernie. 2006. Interview by author, 20 June, Burnaby, BC. Tape recording.

Statistics Canada. 2006. Annual averages of hourly earnings of hourly rated wage-earners, Selected industry groups, Canada, 1945 to 1975. http://www.statcan.

ca/english/freepub/11-516-XIE/sectione/E120_127.csv (accessed 30 April 2006).

Storey, John. 1997. *An introduction to cultural theory and popular culture. Second edition.* Hertfordshire: Prentice Hall/Harvester Wheatsheaf.

Susman, Warren I. 1984. *Culture as history: The transformation of American society in the twentieth century.* New York: Pantheon Books.

Vancouver Sun. 1952. BCCA asks parents to curb young drivers. 1 October, N.p.

Warren, Keith. 2005. Interview by author, 27 September, Tsawwassen, BC. Tape recording.

Chapter 2

Automobile Advertisements:
The Magical and the Mundane

Jim Conley

Since the 1950s, observers on both sides of the Atlantic have viewed the automobile[1] as more than just a means of transportation. For David Riesman and Eric Larrabee (1956) the symbolic meanings of automobiles overwhelmed instrumental meanings. Roland Barthes famously compared the automobile to a Gothic Cathedral, appropriated "in image if not in usage" as a "purely magical object" (1972, 88). Henri Lefebvre likewise declared that the car "is consumed as a sign in addition to its practical use, it is something magical, a denizen from the land of make-believe" (1971, 102). Writing when design departments dominated engineering departments (Flink 1988; Gartman 1994), these authors found magic in the fantastical designs of cars themselves. After designs became more functional in the 1970s, commentators used advertising to explore the cultural meaning of cars. Todd Gitlin maintained that "the car has been a kind of centaur: half conveyance, half fantasy" (1986, 140) while Andrew Wernick argued that cars have been "vehicles for myth" intrinsically bound up with promotional imagery and signs of modernity (1994, 78).

Automobile advertisements are a pervasive feature of contemporary life. Newspapers, magazines, television, the web, billboards, even buses and subway cars display ads for the dominant mode of land transportation. According to estimates made by the industry periodical *Advertising Age*, autos were the most advertised category among their top 200 'megabrands', as 29 brands spent over 5 billion dollars in US media in the first half of 2005 alone (Schumann 2005). What meanings do these advertisements convey? How do they do it? More specifically, to what extent do automobile advertisements draw on mundane or on magical representations? This chapter examines automobile advertisements as a way to understand the dominant place of automobility in culture and society. I argue that automobile advertising assumes and draws on a stock of collective representations and the social organization of automobilized societies. The specific messages conveyed in individual ads refer to well-established representations and depend on them for their meaning. Automobile advertising plays on the opposition between

---

1   Throughout this chapter, 'automobile' will be used to refer to 'light trucks' such as pickups and SUVs as well as passenger cars, and minivans.

automobiles as mundane, albeit highly sophisticated, technical objects, and as magical objects laden with symbolic meanings.

Underlying the question posed in this chapter about the interplay between the mundane and magic is the theoretical debate in automobility literature, including chapters in this volume (Litman and Soron, for example), about the bases for the dominance of automobile travel: does it lie in utilitarian imperatives of travel mandated by the physical layout of urban space – compulsory consumption as Soron calls it – or in the symbolic but no less real realms of status, autonomy, and excitement? Answers to these questions help to untangle how cultural codes sustain the automobile, despite the troubles associated with it, and suggest that strategies to promote alternative forms of mobility need to confront the link between these powerful symbolic representations and automobility.

The analysis that follows suggests that automobile advertising relies on the opposition between the magical and the mundane, the symbolic and the instrumental through operations of distinction and combination. Cars appear in advertising as signs distinguished from and linked to a variety of others, including masculine–feminine,   freedom–constraint,   individual   autonomy–collective dependence, egoism–regulation, desire–discipline, status–dishonour, risk–safety, excitement–luxury, and exterior–interior. Based theoretically on the semiotic analysis of texts (Williamson 1978; Sahlins 1976), this chapter starts from the assumption that ads convey both distinction or differentiation, and combination or association: that is, a given ad will on the one hand distinguish the product it is promoting from competitors, and on the other hand connect it with desires, emotions, or attributes of its audience, drawing on an established repertoire of representations or codes. Although a semiotic analysis presumes that both processes occur simultaneously, it is useful to distinguish them. According to Rappaport's analysis of ritual (1999, in Bellah 2006), lower-level meanings grounded in distinction work through information, middle-order meanings make connections ("similarities, analogies, emotional resonances") through metaphor, and higher-order meanings are "grounded in identity or unity, the radical identification of self with other" (quoted in Bellah 2006, 164). Thus an automobile ad may distinguish a vehicle from others in its class through mundane features (such as safety ratings or equipment, engine size, or length of warranty). Combinations typically involve more powerful associations between the advertised product and emotional states, desires or abstract qualities, such as a Porsche ad [2][2] that calls the Boxster S "Happiness, as purchased with money" and pictures the car on an empty wave-lapped beach, thereby associating it with both an emotional state (happiness) and qualities of natural beauty, solitude, and escape from day-to-day existence. At the same time as advertisements differentiate a vehicle from other vehicles, they promise to distinguish the prospective owner from owners of other vehicles, attributing qualities of each to the other, and totemistically identifying

---

2   Numbers in square brackets indicate the record number of the advertisement in the sample on which this chapter is based.

the buyer with the product (Williamson 1978; Sahlins 1976). The cultural codes behind advertising provide a classificatory schema that both facilitates and constrains how members of the culture think about automobiles, automobility, and themselves. Such a system or cultural code does not determine what people think or say about cars – a code can be used to create many different messages – but it provides the resources for creating those messages, and for understanding them. A semiotic analysis directs us to the necessary decoding work of audiences; for the ads to be meaningful the advertiser and the viewer or reader must share a common code. This approach helps to draw us away from simplistic notions of advertisers manipulating consumers: certainly advertisers try to artfully manipulate the codes to produce interesting messages, to make new distinctions, combinations and identifications, and to draw on consumers' needs, insecurities and anxieties, but as they do so they are constrained by a pre-existing code or set of collective representations. Avoiding the idea that demand is easily manipulated by capitalist producers does not entail accepting the idea of consumer sovereignty: as a cultural process "production is organized to exploit all possible social differentiation by a motivated differentiation of goods" in which "appropriate signs of emergent social distinctions" are developed, objectified and constituted (Sahlins 1976, 184–85).

To begin to understand how automobile advertising constructs meaning, the analysis that follows works on two levels: the unit of analysis is necessarily the individual automobile advertisement, which seeks to sell a particular car model or make; but the objective of the analysis is to examine them collectively, to explore how a corpus of advertisements both reflects and reproduces a cultural code of automobility lending it an aura of inevitability. Both objectives can be approached by identifying themes in what is often called car culture.

Five overlapping representations of automobiles serve as a starting point for investigating the cultural codes of automobility. First, automobility has represented geographical escape, especially in North America (Jasper 2000; Henderson, this volume, on the anti-urban ethos). From the earliest days of mass motorization, cars were associated with escape from unhealthy, congested, dangerous, immoral cities to spacious 'autopian' suburbs, and from industrial civilization to a frontier or to pristine nature that is accessed and mastered by technology (Williamson 1978; McLean, this volume; Sachs 1992). Second, cars have signified social status for individuals or families (Freund and Martin 1993; Packard 1959; Riesman and Larrabee 1956; see Litman's chapter), both when parked in the driveway (Sheller and Urry 2000), and even more when out on the road, where drivers can be known by what they drive, independent of where they live, what kind of work they do, and most other status markers. Third, cars have signified individual freedom from a variety of social constraints (Riesman and Larrabee 1956; Miller 2001): from dependence on other people or organizations and regimentation by their schedules and routes; from face to face interaction with others not of one's choosing, such as occurs on railways and mass transit (Sachs 1992; Henderson, this volume); and from social obligations, especially those associated with civilization and domesticity (Jasper 2000; Inglis 2004; Miller 2001). Cars, as Urry has noted

(Sheller and Urry 2000; Urry 2004; Dennis and Urry, this volume), provide unprecedented levels of individual convenience and autonomy, although this flexibility becomes coercive because once generalized it comes to be expected, even compulsory (see Soron, this volume). Fourth, cars have typically been gendered as masculine, especially when technical knowledge is involved (Freund and Martin 1993), although womb-like interiors have also conveyed femininity (Lupton 1999), so that cars have sometimes been seen as androgynous, with an external masculine side counterposed to an internal, feminine one (Wernick 1994; Miller 2001; Jain 2005). Fifth, cars have also signified opportunities for power, speed and excitement, displaying the driver's skill and daring by taking risks and transgressing norms (Freund and Martin 1993; Genovese, this volume; Sachs 1992).

Unlike most previous studies, that primarily focus on identifying symbolic themes in the cultural representation of automobiles, this chapter investigates relationships between signs, and their frequency, to find the code underlying the advertising messages. This approach reveals the importance of the combination of both mundane or instrumental and magical or symbolic representations of automobiles in advertising. One-sided attention to either the magical or the mundane misses this potent combination.

Looking at Car Advertisements

The method for revealing the underlying cultural codes of automobility used here occupies a middle ground between existing alternatives. Studies of automobile advertising are usually based on either qualitative research of at most a handful of cases (e.g., Gitlin 1986; Jain 2005; Meister 1997), or quantitative analyses of large samples with an applied focus on narrow issues, such as encouragement to speed and neglect of safety (Ferguson et al. 2003; Shin et al. 2005). Instead, like McLean's chapter, this study combines a larger sample size than the former[3] with greater appreciation of complexities of interpretation of meaning – in particular, the polysemic, multilayered, indexical quality of discourse – than the latter (cf. Leiss et al. 1986).[4] To achieve this, I randomly selected one Thursday issue of *The Globe*

---

3   Qualitative accounts have often ignored issues of sampling and representativeness, and in some cases make interpretive leaps from selected cases to large theoretical conclusions. Not all qualitative work suffers from small samples, however: for example Gunster (2004) uses an impressively large sample of print and TV ads for SUVs. Nonetheless, as Shapiro and Markoff show in *Revolutionary demands* (1998), systematic coding produces different results than more impressionistic methods, which tend to focus on what is unusual, rather than the more consistent background messages which for the impressionistic reader become boring, and unnoticed.

4   Traditional content analysis, based on a naive faith in coding instructions and adherence to rituals of quantitative social science, has neglected problems of meaning

and Mail newspaper[5] in each month from February to July 2006, and examined all automobile advertisements from the entire issue, excluding advertisements placed by dealers (because the objective is to analyze the general cultural messages of manufacturers' national advertising campaigns). This procedure produced an initial sample of 154 advertisements; after removing duplicates (that added no new information), 134 ads were left for analysis.[6]

To begin coding, I started with a list of categories adapted from previous content analyses (in particular Ferguson et al. 2003, 827, Table 2.1), and within each, a list of specific features of the message or attributes of the vehicle. Following methods advocated by Shapiro and Markoff (1998) and Franzosi (2004), I coded textual and visual elements of the ads using very detailed categories and descriptions. This detailed analysis is particularly important given the polysemic nature of advertising, and for the same reason, I recorded the tag-line or caption of each ad verbatim, as it often serves to 'fix' other textual and visual messages (see Giaccardi 1995, for a good example). Finally, I categorized the textual content of each ad as either primary or secondary, based on font size and central or peripheral placement in the ad layout (see McLean, this volume, on design in advertising). The analysis to follow focuses mainly on the illustrations, the tag-line, and other primary textual messages. The detailed textual categories are only starting points for what is necessarily an interpretive exercise, and were combined into more theoretically meaningful ones, such as 'excitement', which aggregates power, speed and performance (see below). The frequencies of the resulting distinctions and combinations will be represented graphically. As meaning comes from context, it is important to consider the combinations, oppositions and absences of elements in car ads, and their relative prominence. By showing which signs and combinations of signs appear (and how often, indicated by frequency counts and width of lines connecting elements in each graphic) and which ones do not, the graphics provide a preliminary grammar of automobile advertising.

---

and interpretation. In particular, it has neglected how solutions to problems of ambiguity, double entendre, indexicality, and other complexities of interpretation have relied upon the common-sense background cultural understandings of coders (Franzosi 2004; Cicourel 1964).

5   The issues composing the sample, and the record numbers of the ads in each (in parentheses), including duplicates, were: 9 February 2006 (132–55), 16 March 2006 (109–31), 20 April 2006 (81–108), 18 May 2006 (55–79), 8 June 2006 (1–29), 13 July 2006 (30–54). *The Globe and Mail* was chosen because it claims to be Canada's national newspaper (with different regional editions – the Ontario edition is used here). The Thursday issue contains the Globe Auto section (known until 30 June 2005 as Globe Megawheels), containing road tests and previews of new cars, industry news, racing news, and other auto-related features as well as numerous advertisements. As an 'elite' newspaper, *The Globe and Mail* is biased toward 'high-end' expensive automobiles and the cultural tastes of an educated, high-income audience.

6   Details on the coding categories and sample are available from the author at jconley@trentu.ca.

In what follows, I start by examining lower-order, mundane, instrumental distinctions and symbolic meanings in car ads, then turn to oppositions and combinations involving higher-order symbolic or magical meanings to explore the cultural codes of automobility. The analysis thus starts with messages involving incentives and economy before examining more magical messages that often appear in combination with and opposition to mundane ones: excitement in opposition first to safety, then to luxury; finally concluding with the less frequent considerations of dominance, freedom and gender.

From Instrumental to Symbolic

In Canada, five out of six households own or lease an automobile, truck or van, and private transportation is the largest item of household expenditure after taxes and housing (Statistics Canada 2006). Thus it is not surprising that the most frequent textual messages in the sample of advertisements were mundane and instrumental rather than magical or symbolic. Mundane messages refer to prosaic instrumental features of vehicles, amenable to rational choice based on costs and benefits, and technical and utilitarian considerations, safety, comfort and utility. As Ferguson et al. (2003) found, and Figure 2.2 shows, ads most frequently offered financial

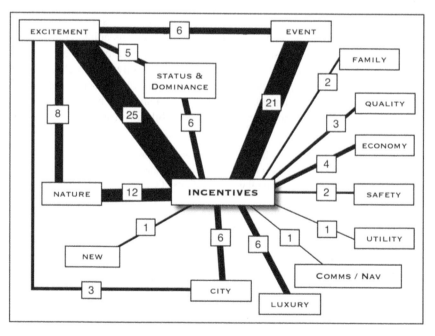

Figure 2.1   Incentives and Related Primary Textual or Visual Elements of Car Ads (N=56)

incentives to purchase or lease a vehicle: special pricing, interest rates, and rebates were primary elements in 56 ads (43 percent) and secondary features of another 49 ads (37 percent). These incentives were often part of temporary sales 'events' such as Honda's "checkered flag event" [8], "The Pontiac red-hot sales event" [9] and Hyundai's "0 percent clearance event" [15], which appeared in 30 ads (22 percent). Since their early twentieth century origins car ads in North America, more than most advertisements, have retained a text-heavy "product information" format emphasizing utilitarian logics focussing on the product itself (Laird 1996; Leiss et al. 1986, 232–4). The prominence of incentives in this sample also reflects the unusually competitive market for automobiles in North America in 2006, spurred by overcapacity and high inventories afflicting General Motors, Ford and Chrysler (Maynard 2006; Saranow and Chon 2006).

Mundane instrumental logic is also apparent in messages of 'economy' in fuel consumption, and in claims of low price or value for money. Fuel consumption is emphasized by a Volkswagen ad: "Dast ist good event. Dast is goodbye to gas stations" [75]; "Over 1,000 bladder busting kilometres per tank" [133]. Despite what were then higher than average gas prices (Keenan 2006), fuel economy appears rarely as a primary message (5 times, 4 percent); more often (25 cases) it is a secondary textual message reporting Transport Canada fuel consumption ratings.

Ads expressed low prices in a variety of ways. A Mitsubishi Lancer ad advises "Stop asking the bank if you can borrow the car" [106]. Mercedes says "We think more shouldn't cost more" [142]. In both cases the subject addressed by

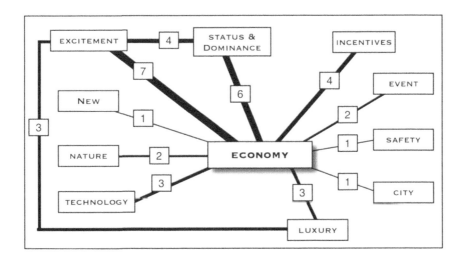

Figure 2.2   Economy and Related Primary Textual or Visual Elements of Car Ads (N=18)

the advertisements is the intelligent consumer, the smart shopper who rationally calculates costs and benefits.

In the information-heavy format of auto advertising, much of the mundane information distinguishing vehicles from competitors appears in the secondary text. Technical features of the engine, steering, drive train or suspension are touted in nearly half of the sample (63 cases, 47 percent), but appear as primary features in only 5 (4 percent). Their distinction is often indicated by being registered or trademarked (in 40 of the 63 cases), as in BMW's "Active Steering" [1] and "Valvetronic engine" [7], Porsche's "horizontally opposed engine" [2], Mercedes' "Direct Control system" [11], Hyundai's "CVVT engine" [15], and Audi's "multitronic CVT" and "FSI direct injection" [124]. Similarly, prosaic messages of utility and convenience appear frequently in the secondary text (50, 37 percent), but are rarely highlighted (3, 2 percent). Roominess, cargo and passenger capacity, manoeuvrability, remote entry, power windows and similar prosaic, lower-level informational meanings that distinguish makes and models are left in the background behind the other mundane messages of incentives and economy, and (as we shall see) more magical messages.

Much consumption of cars is in fact mundane (Carrabine and Longhurst 2002; Soron, this volume). Tempting as it is to underplay the mundane in favour of exploring the more interesting and complex meanings of the car as fantasy object, it is always worth remembering that the purpose of automotive advertising is to sell cars by making claims for 'distinction' in some respect, which for such a major item of expenditure often includes price, financing, and operating costs. More broadly, an emphasis on mundane messages may reflect an ascetic North American ideology, left over from the Protestant ethic, that is susceptible to, but ashamed of appeals to self-indulgence and hedonism: advertising copy that refers to instrumental features of vehicles supplies "vocabularies of motive" (Mills 1940) that furnish defensible utilitarian reasons for a purchase that may have other motivations (see also Marchand 1985). In North America at least, the motor vehicle is thus never just a fantasy object.

By the same token, however, and as the Mercedes tag-line implies, there is 'more': "more power and handling" and "more luxury" [142]. Even when an instrumental logic is on display, more magical messages are frequently conveyed. The car is never just a mundane means of transportation, and as Figures 2.1 and 2.2 also show, ads play on other oppositions and combinations. The Volkswagen ads use humour and the nationality of the manufacturer to convey their fuel economy message, and the Mitsubishi ad sends a message of autonomy as well as economy: when you were young you asked a parent if you could borrow the car; buying a Mitsubishi avoids a similar dependence now.

Figure 2.1 shows that incentives are most frequently combined with visual or textual messages of excitement (aggregating power, speed, and performance), secondarily with sales events, and to a lesser extent with images of nature, and with status and dominance, messages which are themselves tied to excitement. These are represented even in the names of sales events: for example a 'checkered flag event' implies the speed and excitement of automobile racing, reinforced by images of Honda Formula One race cars [8].

As Figure 2.2 shows, messages of economy are most often combined with messages of excitement, status and dominance, and luxury, as well as instrumental incentives. For example, an ad for the Volkswagen Touareg SUV has the tag-line: "Blow your budget" with a script 'e' inserted beneath and between the 'b' and the 'l' to change the word to 'below'; in the secondary text is the statement, "it's enough to make your accountant dance a little jig in his cubicle" [136]. The ad implies that the Touareg is worth blowing your budget on, but does not require it (economy), and contrasts you, the reader, in your big SUV having adventures (perhaps in the Sahara with the vehicle's Berber namesakes) with the dull bean counter cooped up indoors. Similarly, a Volvo ad [13] combines a "limited time offer" event and special lease and finance incentives with a visual representation of the excitement of hard-driving performance in a wild natural setting (Figure 2.3). Thus, while mundane messages are prominent features of car ads, there is more to ads – and to cars – than an instrumental logic. This magical side most frequently involves excitement, combined with and opposed to more mundane messages of safety and luxury.

Figure 2.3   Volvo XC70 2.5T. Reproduced by permission of Volvo Cars of Canada

Source: Globe and Mail 6 June 2006, G8

Excitement or Safety?

Messages of excitement in the form of speed, power and performance are the most frequent primary textual messages after instrumental sales ones (38, 28 percent). In "We build excitement," Gitlin argued that "the car … is the carrier of adrenal energies, a … syringe on wheels" (1986, 138). When visual indications of speed and performance, such as cars pictured travelling on winding roads at speed (indicated, as McLean argues in her chapter, by 'speed blurs' on wheels, on the road or in the background, and sometimes by sand or snow flying up from the wheels, as in Figure 2.3), are included, excitement appears in 65 ads (48.5 percent) (see Figure 2.4). A BMW 323i ad claims "adrenaline-inducing performance" [69]; the Porsche Boxster S promises "the thrill of an open top" and "a drive that let's you know you're alive" [2]; the Acura CSX instructs the reader to "Seize the adrenaline rush" [6]; Mazda supplies your "daily dose of zoom-zoom" and "emotion in motion" [29, 50, 79]; the Audi A4 "rouses the senses" [78]; if you "Need more speed?", the Infiniti M35 advises you to "hold on tight" because it is "breathtaking" and "designed to thrill" [118]; Volvo has "ways to leave you speechless," and is "exhilarating" and "exciting" [151]. An ad for the Porsche Cayman puts technical rationality at the service of masculine virility: beneath the caption "The black t-shirt under the white lab coat," is written "Cold calculating engineering meets pure muscle. Raw strength … flows instantly through a rigid, agile body" [110]. Power is exciting because it is barely under control, as in BMW's "raw power is unleashed" [58]. In both the Porsche and BMW ads, excitement is associated with nature ('pure', 'raw') even when it is based on civilization's advanced technology. In contrast to the mundane informational distinctions considered earlier, here we find metaphor and emotion: these ads associate automobiles with an intense emotional experience that transcends routine daily life.

In contrast to excitement, the message of safety – its mundane opposite – rarely appears in the ads and almost never appears in combination with it. Researchers with an applied interest in safety have argued that "an emphasis on speed and power, without pointing out their deleterious effects, can have the side effect of glamorizing and legitimizing high-speed travel," conveying the message "that speed is fun and risk free" (Ferguson et al. 2003, 830). Although risk may be part of the excitement, the absence of safety in messages of excitement (see Figure 2.4) suggest a strong opposition between the two.[7]

Historically, safety has been a minor theme in car advertisements. The cliché that 'safety doesn't sell' suggests that it would continue to be a minor theme in advertising, but the introduction of new safety features by manufacturers (see Wetmore's chapter), and government (NHTSA) and insurance industry (IIHS) testing of vehicle crashworthiness may have increased its importance. Market research shows that it is an important consideration for potential buyers of their

---

7   In Porsche ads, safety is relegated to the fine print: "Porsche recommends seat belt usage and observance of all traffic laws at all times" [2, 86, 110].

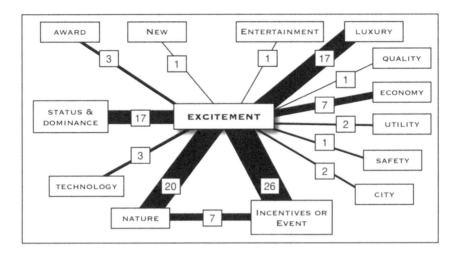

Figure 2.4   Excitement and Related Primary Textual or Visual Elements of Car Ads (N=65)

products, although not in the final purchase decision (Ferguson et al. 2003). Thus we would expect it to appear in motor vehicle advertising, especially if, as MacGregor argues in this volume, automobile manufacturers are now competing in a safety race.

Safety does appear frequently in this sample of car ads, but mostly as a secondary feature. Safety features appeared in 84 ads (63 percent), but safety was a primary element in only 6 (5 percent) of them, the same proportion found for 1993 by Ferguson et al. (2003).[8] Safety is still outnumbered by textual appeals to performance, excitement, luxury and status, and it is almost completely absent as a visual component of advertisements, where speed and performance predominate.

In the six ads where safety is a primary element, occupant protection appears in 4; occupant protection also appears in 42 ads where safety is secondary. An ad for Honda's CR-V and Accord promises that "Honda's commitment to 'safety for everyone' provides outstanding occupant protection" [140]. Performance, in the sense of good handling, braking, nimbleness, and responsiveness, could be considered a safety feature, especially when combined with skilful driving, but as

---

8   In their content analysis of car and minivan television ads from 1983, 1988, 1993, and 1998, Ferguson et al. found that safety was frequently mentioned only in 1993, "when manufacturers were competing to install airbags well ahead of the federal mandate to do so" (2003, 830); see also Wetmore's chapter. In that year safety appeared in 34 percent of TV ads, and was a primary theme in 5 percent. In other years it was a primary theme in 3 percent or fewer of ads, and appeared at all in no more than 8 percent. When Shin et al. (2005) sampled 1998–2002 television ads, they found safety promotion in only 12 percent.

chapters in this volume by MacGregor and Wetmore show, until recently safety advocates in the US have tended to concentrate on passive occupant protection through features such as airbags and crashworthiness which externalize risk onto occupants of smaller vehicles, motorcyclists, cyclists, or pedestrians (White 2004; Gladwell 2001; 2004).

As occupant protection, safety is a passive interior feature that suggests femininity, domesticity and family (Jain 2002). Yet the words 'family' or 'children', or illustrations depicting adults and children together appear rarely in this sample (5 times, 4 percent). In only 2 cases are family and safety combined: Honda promises that the Odyssey "protects your family" (but in the same ad, the Accord "thrills" and the Civic provides "more driving fun") [47]; a Chevrolet ad for the Uplander minivan and Equinox SUV asserts "5-star safety for you and your family" [139]. Explicit connections between safety and gender are entirely absent; in their absence, the relegation of safety to the secondary, mundane, informational text, rather than the primary text or illustrations indicates that occupant protection is subordinate to the message of excitement that is usually coded masculine.

Crash avoidance appears even less often than occupant protection, mentioned only twice in the 6 ads where safety is a primary element; it appears in 58 ads where safety is secondary. The crash avoidance message is often ambiguous: do features such as all-wheel drive signify safety, performance, or something else? An Infiniti ad opposes safety and performance, with the tag-line, "Avoid silp [sic] ups," followed by "all-wheel drive safety when you need it and rear-wheel drive performance when you don't" [143]. All-wheel drive (AWD) signifies domination over nature, especially weather, as in Subaru's "All-Weather Days" (AWD) [105, 121], or Lexus's claim that their AWD gives you "mastery over winter" [153]. An ad for the BMW X3 "sports activity vehicle" SUV [113] mentions AWD, with a graphic that appears to indicate safe driving in snow. Against a snowy plain and cloud-streaked sky, a police officer in fur points a radar gun toward the vehicle, which is turning to the right, its wheels blurred, snow flying up. Safety? Perhaps, but the radar gun and blurred wheels indicate speed, the hard cornering performance, and the tag-line confirms it: "Exhilaration knows no latitude." Northern latitude is implied, but so is an absence of limits, including speed and weather conditions. Safety is subordinate to excitement, and excitement, which is active, exterior and masculine, also trumps comfort and luxury.

Excitement versus Luxury

Ads frequently combine power, speed and performance – excitement – with luxury and comfort as primary elements. Although mere comfort could be considered in opposition to luxury, both are interior to the car, and like safety as occupant protection, they passively shield its inhabitants from the inhospitable world outside. Luxury also implies the good life, one of the central motifs of automobility (Sheller and Urry 2000). In contrast, power, speed and performance point outside the vehicle, to its adventurous passage through that harsh environment.

As Figure 2.5 shows, luxury appears as a primary textual message 21 times (13.6 percent), most often by explicitly extolling "luxury," and otherwise by references to "leather," or wood in the interior, the good life, premium, or refinement. For example, with its tag-line: "Luxury should not be an option," the Lincoln Zephyr boasts leather, wood and luxury [24, 122] and two pages later asserts (in opposition to asceticism), "Depriving yourself of luxury was once considered noble. Now it's just silly" [25, 123]. Acura warns "Consider yourself spoiled" (by a leather interior) [112]; Lexus is "For those who resolve to enjoy life to the fullest" [116]; "the luxurious 2006 Infiniti M" features "an exquisitely appointed leather interior" [111]. Why are leather and wood luxurious? First, they evoke an era of craftsmanship (Gartman 2004): however industrialized the production of car interiors may be, the use of products such as wood and leather, rather than plastic, vinyl, and synthetic textiles, requires attention to the specifics of the natural object (flaws, grain, etc.). It may not be the "British handcrafted interior" of a Bentley [135], but it becomes associated with it. Second, nature is "real" not "artificial," and the ability to afford and appreciate the real thing is status-enhancing.

The implication of luxury in ads is "you deserve it," sometimes explicitly: "You deserve a Lexus. Choose your reward" [153]. The viewer is positioned as a worthy self, both able and entitled to appreciate and enjoy luxury. But not too much.

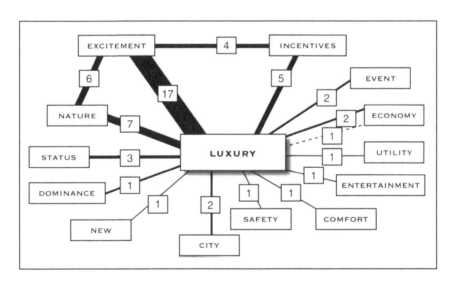

Figure 2.5   Luxury and Related Primary Textual or Visual Elements of Car Ads (N=2)

Figure 2.5 shows that although luxury is combined with status and dominance, and interior features such as comfort and entertainment, the most frequent combinations are with excitement and nature, which are themselves linked in the same ads. Only four ads featuring luxury do not also feature excitement, showing that luxury is not enough; it needs to be exciting too. Although its association with status raises luxury above mundane comfort, both are passive and therefore culturally coded as feminine (Scharff 1991) – performance, power or speed is needed to counteract it. The association of luxury and excitement is shown most strikingly in an Infiniti FX45 ad. The viewer first notices the illustration: A rich chocolate-coloured car body is raised to reveal a bright blue car with racing markings. The tag-line fixes the message: "A luxury SUV with the heart of a sports car" [56]. This ad shows speed and performance in opposition to comfort and luxury at the same time as it combines them. The wolf in sheep's clothing motif of the photo illustration positions the potential purchaser as someone who can passively savour luxury, but still desires active automotive excitement. The association of luxury and excitement is in contrast to the era analysed by Genovese in the previous chapter: once disparaged as dangerous hot rodding, speed and manoeuvrability are now respectable.

An ad campaign for BMW's 5 Series sports sedans also combines luxury and excitement as sporty performance: "So many turns. So many ways to take them" [1], and "A better handling engine?" [51]. The "composure of a luxury sedan" is combined with a "pure sports car in the curves" in the first ad, and its illustration shows an empty road, with tight curves along a rocky, treed hillside; a steering wheel bearing the BMW logo floats above the road at the beginning of a curve. The emotionally satisfying oneness of driver, vehicle and road (Sheller 2004) is visualized here through the absence of most of the car and of the driver; the viewer positioned above and behind the steering wheel can place him or herself behind the wheel, or the engine, figuratively in touch with the road.

Perhaps not surprisingly, luxury is rarely paired with economy (only three occurrences in primary elements), and one of these places them in opposition (indicated by the dashed line in Figure 2.5). An ad for the Infiniti G35 sedan combines expense and distinction in its tag-line: "You practised restraint yesterday" [36]. One of the multiple layers in this message combines past asceticism with present indulgence: you, the reader, tightened your belt in the past, and did without; now you've made it, and you can afford to show it. It also sends a message of desire and freedom from constraint, in opposition to discipline: you no longer have to restrain your passions, and can let go with this powerful 3.5 litre V6, 300 hp, high performance automobile. Luxury's combination with excitement shows that despite its symbolic value, in which the audience can identify with the luxurious attributes of the vehicle as personal qualities, as an interior feature of cars luxury remains to that extent mundane and requires the transcendent, magical invocation of excitement.

Status, Domination and Nature

In nearly one in five ads (26, 19.4 percent) a message of status or domination is conveyed (see Figure 2.6), and as in other symbolic meanings of cars, the audience is invited to identify attributes of the car with themselves in a way that distinguishes them from others. An Infiniti ad characteristically plays on an opposition between mundane transportation and status, with the tag-line "Some cars get you to point B. Others are point B" [12]. By implication, the reader addressed by this ad has arrived in terms of career or status, in contrast to others who are still in the process of getting from A to B. More surprisingly, the mundane category of economy is sometimes combined with a status claim, such as: "Distinction, now in a lower tax bracket" (Hyundai Azera [90]) and "Privilege no longer belongs to the few" (Lincoln Zephyr [149]). To the extent that status depends on associations and high status, and thus on exclusivity (Milner 1994), making the vehicle more accessible appears to dilute its value as a prestige good, but by embracing and overcoming this contradiction the magical in car ads is realized. In what Marchand (1985, 220) called "the parable of the democracy of goods," ads such as this stress both the status boundaries, and the ease of overcoming them.

Since Weber (1968), sociologists have recognized status as a form of power which is often associated with economic power, and messages of dominance sometimes appear in advertisements for luxury brands. A Land Rover ad equates

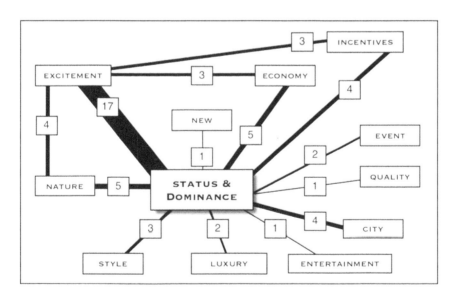

Figure 2.6   Status and Dominance and Related Primary Textual or Visual Elements of Car Ads (N=26)

the vehicle's power with the driver's dominance: "Take a position of power behind the wheel ... it's all at your command ... Make every drive a power trip" [120]. An ad for the Mercedes GL-470 SUV features the tag-line "Own the Road" [55]. The ad shows the vehicle so that the viewer seems to be looking up at it, reinforcing the verbal message of dominance over lesser road users (see McLean, this volume; Gunster 2004).[9] A Jaguar ad with the tag-line "Takes control" superimposed on an illustration of the car suggests that it is the car that exercises power, but an accompanying photo of an elegantly dressed woman looking out from an upper floor window with "gorgeous" written across her body suggests that it is she who is to be controlled. The secondary text advises that the vehicle's technology allows you to "take charge" in "any weather," thus setting up an equivalence between nature and femininity, both of which are to be dominated by masculine technology [150].

Dominance over nature appears frequently in SUV ads (Meister 1997; Gunster 2004; McLean, this volume). In my sample of ads, cars as well as SUVs are frequently shown in wild, natural settings (35 cases, 26 percent), sometimes stationary, sometimes speeding along open roads, and usually with no people or other vehicles in sight. In ads where vehicles are pictured in natural settings, more than two-thirds depict cars only, slightly over one in six depict SUVs only and the remainder show both. These ads do not depict SUVs in natural surroundings more than cars: e.g., Mercedes ads use rugged mountainous terrain as backgrounds for both C-class sedans and GL-class SUVs [53, 64]. Sometimes the relation is even reversed: e.g., an ad for the Ford Mustang and Focus cars and Escape SUV shows the Mustang in a natural setting, the Focus and Escape against blank backdrops [18]. A Volvo ad shows two photos of the XC70 station wagon, one with mountain bikes on the roof, parked next to rocks and a cliff edge, with mountains in the distance; the other shows it in a meadow, with conifers in the background. In contrast, the same ad shows an XC90 SUV in front of an office building, with a man in a business suit walking toward revolving doors, waving to the vehicle's driver [74]. These are examples of reversing the sign (Wernick 1994): when an association, such as between SUVs and nature, is established, it can be switched to convey the magical message: the SUV isn't just rugged, it's urbane and civilized too. This combination of opposites is a central feature of automobile advertising.

Placement of vehicles in natural settings represents them as means of escape from the urban places where most Canadians live (only 13 ads – 10 percent – show vehicles in cities); in the absence of people and of signs of civilization (buildings, road signs, etc.) the driver is freed from social obligations and regulation. The gendered message of geographical escape and social freedom from constraint is evident in some of these illustrations. If a driver is visible, he is male and alone (as in Figure 2.3 above), confirming that escape from civilization to a frontier is gendered masculine. But as McLean finds in her chapter, generally a driver is not

---

9   Dominance can be implied even in ads for more modest vehicles, such as the Mazda6 Touring Edition [79]: "It rules ... by breaking them." Here 'rules' implies dominance, and also excitement through transgressing norms.

visible (as previously noted, Volvo ads such as Figure 2.3 are exceptions). What is the significance of this? On the one hand, to the extent that "masculine" is the dominant category, and therefore unmarked, an absent driver is assumed to be male. As already noted, illustrations showing cars in motion rarely show other vehicles, so this assumed male driver is an individual who as a driver escapes social ties.[10] On the other hand, the absence of drivers (and humans in general) allows advertisers to position the reader as a generic "you": most of the time, the only person "in" the ad is outside it – "you" the reader.[11] Because no one is in the ad, "you" can be anyone and "you" can identify yourself with the vehicle, so that its distinction from other vehicles becomes your distinction from other people, and the transcendental experiences it promises can be yours if you visit "your dealer." The absence of people in most ads thus represents freedom from social exclusions, especially as the "you" being addressed is an unattached individual. Even when status is being sold, the logic of the market-driven advertising is not exclusionary. Although collectively automobile advertisements sell automobility, individually their objective is to sell cars; nothing is to be gained by excluding categories of people. Thus the few ads that include people who are not driving often show people of different colours, and both males and females (e.g., Honda [114, 140], Chrysler [148]). As Sahlins (1976, 184) argued, the symbolic code in capitalist society is "an open set, responsive to events" and I would add, to social trends, assimilating and objectifying them. Thus, for example, with the exception of the Jaguar ad discussed above, none of the ads in this sample position women as automotive accessories. The symbolic code of automobiles and automobility is not closed and deterministic, and we must be careful not to confuse cultural representations with social exclusions. The cultural codes of automobility are related to social classifications, but the latter are not mapped directly onto the former. Part of the magic in automobile advertising is escape from social constraints, including social obligations and categorical class, gender, and racial inequalities (Riesman and Larrabee 1956; Miller 2001).

---

10   In this sense, even ads without people can present "social tableaux" defined by Marchand (1985, 165) as "advertisements in which persons are depicted in such a way as to suggest their relationships to each other or to a larger social structure." The solitary or absent, implicitly masculine driver indicates social relations of an individualistic, even egoistic character.

11   The reader is directly addressed using "you" or "your" in 65 ads (49 percent) and indirectly in the imperative in another 17 ads (13 percent). Another 29 ads (22 percent) referred to "your dealer." In only 22 ads (16 percent) was the reader not addressed in the second person in some form.

Conclusion

The magical in automobile advertising appears elsewhere as well in the combination of an urban built environment with natural features, portrayed in three ads. In the Infiniti ad mentioned above [56], the windows of a modernist building reflect the open countryside. The sole Toyota ad in the sample shows a Camry in front of the drive-through menu for "Marcel's Gourmet Express," in the distance, across empty fields, are the buildings of a city. The caption reads "What you want is what you need" followed in the secondary text by "Do you want best quality or best value? How about both?" [3].

The paradoxical message of gourmet fast food is a metaphor for the main message of automotive advertising, that you can have both: both urban and natural amenities, both luxury or safety and excitement, both economy and status, both the mundane and the magical. This union of opposites is part of the appeal of auto-mobility, on which advertisements draw through operations of distinction, combination and identification.

This study of automobile advertising suggests that the cultural code on which car advertisements play can be usefully organized around the opposition between the magical and the mundane. It suggests that however compulsory the consumption of automobiles may be (Soron, this volume), its meaning goes beyond instrumental transportation to include the more symbolic attachments and identities of the abstract desire of consumer capitalism (Milner 2004).

This analysis also suggests possibilities for future research. First, understanding the cultural codes on which car ads draw would benefit from the study of different media, such as magazines, the web and television. This would serve both to correct for possible biases of this sample (including over-representation of expensive vehicles, and perhaps of mundane messages in daily newspaper advertising), and to investigate if and how the codes of automobility vary by audience. For similar reasons, historical and comparative research on automobile advertising would show if and how codes (not just messages) vary by cultural and historical environment, or if they are constant, reflecting inherent features of mass automobility.[12] Methodologically, this chapter has tried to show that such research should be undertaken with a sample size large enough to reveal patterns, but small enough to permit attention to multiple levels of meaning. For that matter, there is no reason that such research should be restricted to advertising: literature, journalism, films, and television programs are also points of access to the cultural codes of automobility. If we are to understand the place of the automobile in societies, we need to consider it in all its dimensions, including its symbolic meanings as well as its instrumental uses. If we are to challenge the dominance of automobile transportation, we need to understand the operation of cultural codes that support

---

12   Previous research suggests intriguing differences between North America and Europe (Paterson and Dalby 2006; Paterson 2007), and within the latter, between Britain and Italy (Giaccardi 1995).

it, and learn to use them to challenge it.[13] In the following chapter, McLean shows how environmentalist critics of SUVs have attempted to do this.

References

Barthes, Roland. [1957] 1972. The new Citroën. In *Mythologies*, Annette Lavers, trans., pp. 88–90. New York: Noonday Press.

Bellah, Robert. [2005] 2006. Durkheim and ritual. In *The Robert Bellah reader*, Robert N. Bellah and Stephen M. Tipton, eds, pp. 150–80. Durham: Duke University Press.

Carrabine, Eamonn and Longhurst, Brian. 2002. Consuming the car: Anticipation, use and meaning in contemporary youth culture. *Sociological Review* 50: 181–96.

Cicourel, Aaron V. 1964. *Method and measurement in sociology*. New York: Free Press.

Ferguson, Susan A., Hardy, Andrew P. and Williams, Allan F. 2003. Content analysis of television advertising for cars and minivans: 1983–1998. *Accident Analysis and Prevention* 35: 825–31.

Flink, James J. 1988. *The automobile age*. Cambridge: MIT Press.

Franzosi, Robert. 2004. *From words to numbers: Narrative, data, and social science*. Cambridge: Cambridge University Press.

Freund, Peter and Martin, George. 1993. *The ecology of the automobile*. Montreal: Black Rose Books.

Gartman, David. 1994. *Auto opium: A social history of American automobile design*. London: Routledge.

———. 2004. The cultural logics of the car. *Theory, Culture and Society* 21: 169–95.

Giaccardi, Chiara. 1995. Television advertising and the representation of social reality: A comparative study. *Theory, Culture and Society* 12: 109–31.

Gitlin, Todd. 1986. We build excitement. In *Watching television*, Todd Gitlin, ed., pp. 136–61. New York: Pantheon.

Gladwell, Malcolm. 2001. Wrong turn: How the fight to make America's highways safer went off course. *New Yorker* 11 June: 50–61.

———. 2004. Big and bad: How the SUV ran over automotive safety. *New Yorker* 12 January: 28–33.

Gunster, Shane. 2004. 'You belong outside': Advertising, nature and the SUV. *Ethics and the Environment* 9(2): 4–32.

Inglis, David. 2004. Auto couture: Thinking the car in post-war France. *Theory, Culture and Society* 21: 197–219.

---

13 For example, it would be worthwhile to explore how some of the cultural codes examined here might help render public transportation a 'cool' form of mobility (see Litman's chapter for a discussion about redefining the value of public transportation).

Jain, Sarah S. Lochlann. 2002. Urban errands: The means of mobility. *Journal of Consumer Culture* 2: 385–404.

——. 2005. Violent submission: Gendered automobility. *Cultural Critique* 61: 186–214.

Jasper, James M. 2000. *Restless nation: Starting over in America.* Chicago: University of Chicago Press.

Keenan, Greg. 2006. Auto makers eye fuel economy. *Globe and Mail* 24 May. http://www.theglobeandmail.com/servlet/story/RTGAM.20060524.wxr-autos25/BNStory/Business/ (accessed 26 May 2006).

Laird, Pamela. 1996. 'The car without a single weakness': Early automobile advertising. *Technology and Culture* 37(4): 796–812.

Lefebvre, Henri. [1968] 1971. *Everyday life in the modern world,* Sacha Rabinovitch, trans. New York: Harper Torchbooks.

Leiss, William, Jhally, Sut and Kline, Stephen. 1986. *Social communication in advertising: Persons, products, and images of well-being.* Toronto: Methuen.

Lupton, Deborah. 1999. Monsters in metal cocoons: 'Road rage' and cyborg bodies. *Body and Society* 5: 57–72.

Marchand, Roland. 1985. *Advertising the American dream: Making way for modernity, 1920–1940.* Berkeley: University of California Press.

Maynard, Micheline. 2006. Detroit flails in latest effort to reinvent itself. *New York Times* 16 September. http://www.nytimes.com/ (accessed 17 September 2006).

Meister, Mark. 1997. 'Sustainable development' in visual imagery: Rhetorical function in the Jeep Cherokee. *Communication Quarterly* 45: 223–34.

Miller, Daniel. 2001. Driven societies. In *Car cultures,* Daniel Miller, ed., pp. 1–33. Oxford: Berg.

Mills, C. Wright. 1940. Situated actions and vocabularies of motive. *American Sociological Review* 5: 904–13.

Milner, Murray Jr. 1994. *Status and sacredness: A general theory of status relations and an analysis of Indian culture.* New York: Oxford University Press.

——. 2004. *Freaks, geeks, and cool kids: American teenagers, schools, and the culture of consumption.* New York: Routledge.

Packard, Vance. 1959. *The status seekers.* New York: David McKay.

Paterson, Matthew. 2007. *Automobile politics: Ecology and cultural political economy.* Cambridge: Cambridge University Press.

Paterson, Matthew and Dalby, Simon. 2006. Empire's ecological tyreprints. *Environmental Politics* 15(1): 1–22.

Rappaport, Roy A. 1999. *Ritual and religion in the making of humanity.* Cambridge: Cambridge University Press.

Riesman, David and Larrabee, Eric. 1956. Autos in America. In *Abundance for what? and other essays,* David Riesman, ed., pp. 270–99. Garden City: Doubleday.

Sachs, Wolfgang. [1984] 1992. *For love of the automobile: Looking back into the history of our desires,* Don Reneau, trans. Berkeley: University of California Press.

Sahlins, Marshall. 1976. *Culture and practical reason.* Chicago: University of Chicago Press.

Saranow, Jennifer and Chon, Gina. 2006. Car dealers keep discounts rolling. *Wall Street Journal* 3 August.

Scharff, Virginia. 1991. *Taking the wheel: Women and the coming of the motor age.* New York: Free Press.

Schumann, Mark. 2005. Mega brands. *Advertising Age* 76: 41.

Shapiro, Gilbert and Markoff, John. 1998. *Revolutionary demands: A content analysis of the cahiers de doléances of 1789.* Stanford: Stanford University Press.

Sheller, Mimi. 2004. Automotive emotions: Feeling the car. *Theory, Culture and Society* 21: 221–42.

Sheller, Mimi and Urry, John. 2000. The city and the car. *International Journal of Urban and Regional Research* 24: 737–57.

Shin, Phillip C., Hallett, David, Chipman, Mary L., Tator, Charles and Granton, John T. 2005. Unsafe driving in North American automobile commercials. *Journal of Public Health.* 27, 4: 318–25.

Statistics Canada. 2006. Spending patterns in Canada 2005. Income Statistics Division. Ottawa: Minister of Industry. Catalogue no. 62-202-XIE.

Urry, John. 2004. The 'system' of automobility. *Theory, Culture and Society* 21: 25–39.

Weber, Max. 1968. *Economy and society: An outline of interpretive sociology.* Guenther Roth and Claus Wittich, eds, Berkeley: University of California Press.

Wernick, Andrew. 1994. Vehicles for myth: The shifting image of the modern car. In *Signs of life in the U.S.A.: Readings on popular culture for writers,* Sonia Maasik and Jack Solomon, eds, pp. 78–94. Boston: Bedford Books.

White, Michelle J. 2004. The 'arms race' on American roads: The effect of sport utility vehicles and pickup trucks on traffic safety. *Journal of Law and Economics* 47: 333–55.

Williamson, Judith. 1978. *Decoding advertisements: Ideology and meaning in advertising.* London: Boyars.

Chapter 3

SUV Advertising: Constructing Identities and Practices

Fiona McLean

Despite mounting concerns from various groups over the social and environmental impacts of the Sport Utility Vehicle (SUV),[1] it has had phenomenal market growth in the last couple of decades. From the mid-1980s until recently SUVs were the fastest growing category in motor vehicle sales and were credited with keeping the declining North American auto industry profitable (Gunster 2004). In the US, where these vehicles have been most popular, SUVs accounted for only 1.8 percent of new vehicle sales in 1980, but by 2002 their market share had grown to over 25 percent. The greatest increases during that period were between 1995 and 2002 when the number of SUV models available almost tripled from 28 to 75 (Gunster 2004). While the UK has significantly fewer of these vehicles than the US, sales have increased year on year, and European carmakers such as Renault, Saab and even Alfa Romeo have developed new four-wheel-drive models.

Campbell credits the boom in 4x4s with their military heritage, as well as with the opportunity they offer to "express a rugged individualism" and to "feel a bond with the great outdoors and the American frontier" (2005, 957). He argues that growing militarization, remasculinization, dependency on oil, and domestic and international insecurities underpin a radically individualized citizenship, which Mitchell (2005) refers to as the 'SUV model of citizenship'. Yet, paradoxes abound with the popularity of the SUV (Campbell 2005). Despite associations with 'the great outdoors', statistics in the UK have shown that only 12 percent of 4x4 owners use their vehicles off-road, while 40 percent never leave the city (Macfarlane 2005). Moreover, the SUV has become an urban luxury vehicle with increasingly plush interiors and an extensive array of technological gadgets. Combining elements of style and comfort with functionality and practicality, the SUV "can uniquely fuse the hitherto 'uncool' aspects of family life with the hipness of the outdoor adventure (forging new routes through both). In that sense, they have appealed to young families who want to associate and be associated more with the outdoors and adventure than with the suburbs and errand running" (Jain 2002, 398). Where once the minivan was the most popular family wagon,

---

1   I have included all models with four-wheel-drive capabilities in the category of SUVs (not including pick-up trucks) and have used the terms SUV, 4x4, and four-wheel-drive vehicle interchangeably.

it was pushed out of this spot as more and more families adopted SUVs as their vehicle of choice (Bradsher 2002).

This chapter is based on the results of an examination of SUV advertisements in the UK from January 2004 to September 2005. After selecting a variety of monthly magazines (see Appendix for the list), I recorded all the SUV ads along with brief descriptions, and then chose 77 of them to analyze fully. I used semiotic analysis[2] to interpret the meanings and connotations in each of the ads individually and then grouped them together to identify commonalities and differences in the themes, cultural codes and compositional elements. In what follows, I first discuss the significance of advertising in shaping consumer demand. Second I examine specific design and photographic techniques – largely neglected by the social sciences – that compose and construct advertising images in SUV ads. Third, I focus on the advertising settings of SUV ads and discuss the oppositional themes of 'nature' and the 'urban jungle'. I then conclude with a discussion of the anti-SUV movement and the next generation of ads.

Advertising: The Shaping of Demand

Almost 100 years ago, Walter Dill Scott claimed that "advertising has as its one function the influencing of human minds. Unless it does this it is useless and destructive to the firms attempting it" (Scott 1908, 2 cited in Craig 1992, 167). Since then, advertising has become a multi-billion dollar industry, with large portions of promotional budgets spent on it every year. In the case of SUVs, advertising expenditures in the US alone increased from an estimated $172.5 million in 1990 to $1.5 billion in 2000 (Bradsher 2002). The goal of advertising is to create an urgent desire in potential consumers by developing a sense of lack and feelings of inadequacy (Bordwell 2002, 242). Symbolic meanings are attached to goods and their properties by associating them with our goals, values and dreams. The implicit suggestion is that owning or purchasing the product is the route to happiness and success.

Stokes and Hallett (1992) suggest that because advertising produces media-styled norms of behaviour and taste, it plays a key role in the emotional attachments people form towards products, such as their cars. One of the ways that ads do this is by making cultural and psychological appeals which attempt to construct an identity and lifestyle for the potential buyer (Wernick 1991; Williamson 1978). However, Williamson argues that people "do not simply buy the product in order to become a part of the group it represents; [they] must feel that [they] already, naturally, belong to that group and therefore [they] will buy it" (1978, 47).

Although advertising is an important source of meaning, it is just one part of a "wider process of cultural commodification" (Wernick 1991, 181). Nevertheless,

---

2   Semiotics examines signs and considers how meanings in images or texts are constructed.

many critics believe that it has too much power and influence over the circulation and distribution of ideological values and that together with consumption it is a "powerful socializing force that feed[s] on various cultural sites to establish the meaningfulness of products" (Craig 1992, 177). For example, Litman (this volume) claims that the vehicles people choose, or even their decision to drive, is influenced by 'positional value', which is often portrayed in advertising. As a result, consumers choose more expensive, higher-value models than they otherwise would. At the same time, public transport or smaller, more fuel-efficient models become stigmatized because they lack status.

While advertising is not the only influence in people's lives, and perhaps not even the main one, it does reinforce people's prior dispositions, as well as other cultural, social and economic factors (Dyer 1982; Stokes and Hallett 1992). However, advertising cannot dictate consumer behaviour, nor is it able to create desire for a product that doesn't appeal to consumers. Furthermore, individuals may assign different meanings to the same product and may not interpret an ad in the way originally intended by the designers (Heffner et al. 2006). Nonetheless, advertising sustains the "more general cultural, behavioural, ideological and ethical principles which we apply in our everyday social activities" (Dyer 1982, 136). It offers a particular structure for interpreting the world and helps to create an ideological framework, that is, the system of values, beliefs, norms and behaviours that fit in with people's self-identities and guide their choices and decisions (Dyer 1982; Heffner et al. 2006; Williamson 1978). In particular, Heffner et al. assert that the widespread adoption of SUVs has been influenced to some degree by "consumers' enthusiastic response to the symbolic meanings" of these vehicles (2006, 1). Design techniques are a critical element in creating the symbolic meanings of these advertisements.

The Design of Advertising Images

The social sciences have largely overlooked the construction and composition of images in favour of the study and interpretation of objects and signs (Whitely 1999 in Rose 2001), despite the fact that the way an audience views an image is highly dependent on photographic techniques and the formal arrangement of visual elements in the picture (Dyer 1982; Rose 2001). A photograph, for example, is "composed not just in the usual sense by the photographer but by conventions of colour, lighting and subject which [help] to fix meaning" (Geraghty 2000, 363). McQuarrie and Mick claim that "it is not possible to change the style of an advertisement without also changing some of its meaning" (1999, 38). Design features are constantly interacting with shifts in cultural codes by (re)organizing dominant meaning patterns and reconciling the core values and symbols that have currency (Dyer 1982; Wernick 1991). Advertising designers use specific photographic techniques to shape the meanings in ads – physical location and setting; colour; framing (angle, proximity and spatial organization); movement; image manipulation; text; and people.

There are four main physical settings in SUV ads (McLean 2005): a nature or wilderness setting; an urban milieu; a location that is neither specifically in the wilderness nor in an urban environment (e.g., at a gas station or outside a sports stadium); or a plain white or coloured studio backdrop. These settings show viewers and potential buyers how and where the vehicles should be used, thereby influencing people's experiences of the places and activities advertised and shaping the values and practices attributed to SUVs (Heffner et al. 2006; Wilson 1991). Advertisements for 4x4s tend to use 'masculine' colours in the images: silvers, greys, browns and blacks. Tones are often muted, shades are pale or washed out, and the colour saturation is low. Most ads display silver vehicles which "accentuate edgier designs and convey an aura of technology" (DuPont 2006, 2), while a black vehicle implies aggression and a desire to command respect (IOL 2007).

The framing (angle, proximity and spatial organization) of the image adds much to the meaning of an advertisement. The angle from which the object is photographed will guide the viewer's eyes to the image in a particular way. For example, vehicles shot from a low angle convey superiority, status or symbolic power, while images shot from a high angle, that is, above the object, give the viewer maximum power (Kress and van Leeuwen 2006). This latter technique is common in many SUV ads set in nature, placing the viewer "above and dominant over a static and subordinate landscape, which lies out beyond us inert and inviting our inspection. Such photographic practices thus demonstrate how the environment is to be viewed, dominated by humans and subject to their possessive mastery" (Urry 1999, 4). Shooting a vehicle from straight on, especially when in close range, can be interpreted as confrontational and considered as a means of intimidation and subordination (Messaris 1997). The majority of SUV ads are shot so that the vehicle is positioned obliquely at a ten to forty-five degree angle, allowing the viewer to more thoroughly examine features of both the front and the side.

Proximity describes the distance of the object from the viewer and affects the level of attention and engagement. Messaris (1992) asserts that the closer the viewer is to the object, the more s/he will identify with it and the more it will facilitate an imaginary interaction. However, tight close-ups of vehicles, particularly those from a low frontal angle, can be experienced as an act of aggression and hostility and may produce feelings of claustrophobic intensity (Kress and van Leeuwen 2006; Messaris 1997; Rose 2001). Long shots combined with a high angle are frequently used in SUV ads with connotations of domination over the landscape, thereby allowing viewers to "contemplate the world from a god-like point of view, put[ting] it at [their] feet" (Kress and van Leeuwen 2006, 145).

Spatial organization defines how designers arrange objects within an image. One motif is for the vehicle to dominate the image, which allows greater viewer identification. Another common layout places the vehicle 'in motion' at the right side of the page, facing to the right, and often in profile. The layout depicts the vehicle heading into the unknown of the uncharted territory off the edge of the page, after having travelled across the landscape in the image, thereby proving its capabilities and its mastery of the terrain. Stationary vehicles often appear on the left side of the page (again facing to the right), poised for action, ready for anything, and set to drive 'across the page'.

Movement is another major photographic technique in automobile advertising. A photograph of a moving object will appear in one of two ways: the object will be in focus with the background blurred, or the background will be in focus with the object blurred. The majority of ads that depict the SUV as moving show both the vehicle and the background in clear focus, which means the photographer actually shot the SUV while it was stationary. The illusion of movement is added through digital manipulation to give an impression of speed and demonstrate the prowess of the vehicle. Other digital alterations in the ads include erasing shadows, brightening colours, darkening windshields or adding reflections, while more extreme cases involve inserting the vehicle into an existing image or using computers to generate an imaginary landscape.

The accompanying text in an advertisement is also a design technique that helps to anchor the meaning. As McQuarrie and Mick point out, "text structure tends to shape or direct consumer response, even though it does not, strictly speaking, fix or determine it" (1999, 38). Most SUV ads follow a standard formula: a tagline consisting of a phrase or a couple of sentences in a large bold font, usually in the top left of the image; a box or an area in the image with smaller text which gives the name of the vehicle and information about its features; and tiny font at the bottom of the page outside the frame of the ad listing the fuel consumption figures and CO_2 emissions ratings.[3] An exception to this is in Land Rover ads which frequently have no text at all other than the logo and name of the model. When a product becomes a signifier in its own right, that is, a product goes from representing a quality or lifestyle to *being* that quality or lifestyle, "it may become not only 'sign' but the actual *referent* of that sign" (Williamson 1978, 36). The Land Rover brand has therefore become so recognizable that it speaks for itself and the meaning has become so embedded that it no longer needs text (McLean 2005).

Design techniques also include the presence or absence of people and gender connotations. The exclusion of people in advertisements is very common and invites viewers to insert themselves in the image as the 'leading actor' (Williamson 1978). In SUV ads, the absence of people is striking. When people are present, there is usually only a solitary male driver. Typically he cannot be clearly distinguished because of tinted windows or reflections and shadows on the windshield. This muted presence is regularly the case in ads where the vehicle is in motion, whereas those with stationary vehicles tend to have no people. Where people are absent from the ad, there is usually an implied male subject (Jain 2005; Paterson and

---

3   This format of the text for fuel consumption/CO_2 emissions actually contravenes EU Directive 1999/94/CE which states that the information giving the vehicle's fuel consumption and CO_2 emissions figures should be easy to read and no less prominent than the main part of the information in the advertisement (see Advertise CO_2).

Dalby, 2006). The outdoors has largely been identified with masculine qualities and interests (Wilson 1991) such as danger, excitement and adventure. Therefore, since designers frequently set SUV ads in the outdoors where rugged individualism and the 'conquering of the frontier' are key themes, they are promoting masculine traits and perpetuating the idea of men as the main drivers of these vehicles. Urban-set ads, on the other hand, tend to show people more often than those set in nature, thereby reinforcing the idea that escaping to the outdoors will mean going to a setting devoid of any human presence – that is, nature is best appreciated "on its own" (Wilson 1991, 36).

Despite the increasing number of women who are buying cars, sexist portrayals of gender roles in vehicle advertising are still commonplace (Paterson 2007). For example, the tagline "Comes with map reader minus the argument" for a Honda CR-V ad run in early 2005 points to the stereotype that women can't read maps and men won't ask for directions and suggests that arguments will be avoided thanks to the vehicle's built-in satellite navigation system. In an SUV ad that is specifically targeted to a female audience, stereotypes are further reinforced. The foreground is dominated by the blurred image of a woman's shoe covering the left page and the bottom of the right. The rest of the right page shows a woman stopped at an intersection on a London street, leaning out of the window of her Jeep Grand Cherokee to get a better view of the shoe. The main tagline says "£350 for shoes you can't walk anywhere in? But then you won't have to, will you?" The text below continues with "Who needs sensible footwear when a Jeep Grand Cherokee will get you from A to just about anywhere? And with all the luxury you could ask for, and more capability than you're likely to need, what could be a more practical buy?" Many car ads aimed at women, such as this one, promote the vehicle as nothing more than a fashion accessory and perpetuate stereotypes that men use cars for important matters while women's driving is not only less important (Paterson 2007), but even frivolous. The design of cars themselves is also gender oriented, either constructing masculinity as dominance, with the car being a (sexual) extension of the male, or femininity as submission, with the car being equivalent to a mistress or wife (Paterson 2007; Wernick 1991). While many sports cars and luxury models are cast in the role of a woman, SUVs are striking in their gender ambiguity. The tough image and manifestation of masculine power preclude them from being feminized, while at the same time through their womb-like properties they are the embodiment of the maternal protector (McLean 2005; Wernick 1991; see also Conley, this volume).

A Land Rover Discovery ad (*Men's Health* 2005) illustrates a number of the photographic design techniques, discussed above, through a visual narrative presented as an 'adventure story'. The computer-generated image seamlessly stitches together a range of diverse environments from the towering office blocks of the urban centre to the ice-capped mountains, the arid desert canyons and the dormant birch forests. The vehicle has seemingly journeyed through a variety of "external disconnected landscapes that are there for no other purpose than to be subdued as a clear illustration of the prowess of the driver" (Paterson and Dalby

2006, 5). Low saturation, muted colours and dark shades accentuate the rawness of nature. The angle of the shot from below the bumper shows the vehicle's power, while its proximity enables the viewer to identify with it. The placement at the right side of the page implies that it has come from the city in the back left of the image, having travelled across a range of terrains, conquering each of them. Apart from the digital creation of the image itself, other manipulations include the reflection of the canyon walls (which are outside the frame of the image) in the vehicle's doors, as well as the blacked out windshield and side windows. As is common with other Land Rover ads, there is an absence of both text and people. The ad implies a male subject with references of boldly traversing harsh landscapes, powerfully overcoming all obstacles and skilfully mastering the technology that allows him to conquer nature.

Nature: 'The Call of the Wild'

The physical location and the environmental setting create the context in which the meanings of ads can be interpreted. I therefore focus here on the two most common settings used in SUV advertising, nature and the 'urban jungle', and show how they 'sell' the vehicle. Nature is the primary referent of a culture. It is the 'raw material'" (Williamson 1978, 103) within a signifying system which transforms the meanings of objects and through which "a social order is communicated, reproduced, experienced, and explored" (Williams 1982, 13 in Meister 1997, 227). Through the articulation and adaptation of established cultural codes and understandings, advertising communicates appropriate and acceptable ways of consuming nature and the environment (Hansen 2002).

SUV advertisements regularly use imagery of nature and the wilderness as a central theme to sell the vehicle. By placing products in nature, advertisers endow them with qualities of 'the natural' (Williamson 1978) not only in the sense of belonging to the natural world, but also of possessing its intrinsic attributes. In addition, placing SUVs in natural surroundings shows them as being put to the obvious, 'natural' use for which they were intended.

Driving has long been an activity that allows people to leave the city and suburbs in order to venture out to the wilderness or countryside. Escaping the confines of a city plagued with congestion and pollution is therefore the 'natural' response to life within an urban civilization, and the car provides the means for this escape. Indeed, "the very idea of 'nature' that many anti-car campaigners are defending may have been constituted largely through automobility" by developing "contemporary appreciation for the extensive, relatively untouched and visually pleasing vistas that environmental campaigners seek to preserve" (Sheller 2004, 231).

Advertisers cast nature in a number of ways: as a vulnerable and fragile environment that should be protected and safeguarded; as a pristine, unspoiled paradise; as an in-between place or obstacle to be overcome on the way to a final destination; as a place of freedom and rugged individualism where one can challenge oneself, test one's mettle and prove one's valour; as a place to be improved, shaped or controlled, particularly through technological means; and, finally, as a dangerous, hostile, and unpredictable enemy to be conquered (Hansen 2002; Williamson 1978).

While the image of nature as fragile and vulnerable is a central theme in news stories or documentaries, in his analysis of UK television advertising, Hansen (2002) found that virtually no ads showed this message; only 3.8 percent of the advertisements in his study were based on 'nature as an enemy'. In contrast, my analysis of SUV advertising shows this latter theme to be prominent. Nature as a proving ground and a place to be dominated and improved upon are also common representations in SUV ads.

A Volkswagen Touareg ad (*Men's Health* 2004) casts nature as a threatening, forbidding environment. The landscape in the image is barren, unpopulated and desolate. The background lighting is weak, indicating that night is approaching. Clouds overhead suggest that a storm is brewing and nasty weather is moving in. Snow-covered mountains in the distance indicate that it is a cold time of year. Jagged rocky outcrops, boulders and land without vegetation give the impression of an inhospitable terrain. The foreground is dark and ominous, giving a sense that the vehicle is driving into an even more hostile and unwelcoming territory. The nature in this image has connotations of risk and danger, while the Touareg signifies safety and security. The vehicle provides not only the means to escape the oppression of civilization but also the means to get away from the threats and hazards posed by nature and return safely to civilization (Holder 1991; Williamson 1978). The Touareg is therefore endowed with the capability to fend off danger while providing excitement and adventure.

In another Touareg advertisement (*Car* 2005), the designers address the need for creature comforts by focusing on the 'inner' experience. The vehicle is facing upwards on a hill at a thirty degree angle in another dark, barren setting. The tagline, which reads "Leather seats, sat-nav and 6 CD autochanger now standard. (Ah, the great indoors.)," sets up the contrast between the opposing elements of nature and civilization. As Jim Conley (this volume) points out, it is a common practice in car advertising to play on the opposition of 'binary codes' by contrasting the interior of the vehicle with what is exterior to it. In this case, the meaning is that nature can be improved through technological means and the vehicle itself represents a wrap-around experience. The ad suggests that the experience of being in the vehicle is itself an escape, therefore making a trip to the outdoors almost superfluous. Again the inside of the Touareg provides a safe haven, protecting the occupant(s) from the exterior risks and threats presented by nature.

The message in both ads is not only that the Touareg can handle harsh terrain or extreme environmental conditions, but that it is endowed with the power to overcome and defy nature. The SUV, then, is not just another item of equipment or an accessory for an outdoorsy, active lifestyle; it provides the technological means for the domination of nature. However, it is not enough to dominate nature; it must also be civilized (Paterson and Dalby 2006). Once more, the vehicle is the instrument through which this goal can be achieved. With its luxurious interior and refined features, the SUV is the embodiment of 'portable civilization' (Garnar 2000), enabling the occupant(s) to bring along all the home comforts.

SUV ads in the UK rarely depict landscapes found in Britain. Instead, they show images of foreign or exotic destinations with mountains, deserts and canyons. Paterson and Dalby claim that this portrayal reveals imperial attitudes towards the landscape and has a "specifically American mode of understanding space, nature, power, territory, etc." (2006, 15). This portrayal fits in with the myth of the frontier: the conquest and expansion of new territory, subduing native peoples and the appropriation and occupation of the wild land in order to 'civilize' it.

The Urban Jungle

As with many of the nature-set ads, SUV advertisements with an urban background also emphasize the notion of 'portable civilization', showing the vehicle as a mobile dwelling or a cocoon on wheels. Another central theme, however, portrays the SUV as an urban assault vehicle (Garnar 2000). Both these types of ads depict the city as an unfriendly and uncivilized environment from which one needs to escape. They also highlight the technology systems within the vehicle which facilitate this escape.

Wernick notes that representations of vehicles do not show them as "what [they] overwhelmingly [are]: a commuter vehicle tail-gating traffic to and from work"; instead the vehicles signify "a means to get away – a piece of magic, in fact, which could wash away toil and traffic altogether" (1991, 78). Vehicle advertisements often give the impression that urban centres are devoid of both cars and people, magically (as emphasized in Conley, this volume) erasing the infrastructure which supports mass automobility (Gunster 2004). This absence can be linked to the increasing concentration on privatism over the past few decades, which highlights individual consumption, subjective experiences, personal space, and the right to be left alone (Campbell 2005; Gunster 2004; Mitchell 2005; Wernick 1991).

Ads which focus on the idea of SUV as cocoon emphasize the 'inner' experience. Urry suggests, "the worse the roads outside the greater the pleasure, security and sense of dwellingness that is sought within the car" (2000, 6). Simulating the comforts of the domestic environment, the vehicle becomes a living room or entertainment centre on wheels with reclining leather armchairs, a stereo system and drinks holders. Along with "incessant celebrations of a luxurious interior" (Gunster 2004, 23), which have a long history in car advertising, the ads also highlight the feelings invoked by the vehicle's internal features – the driver will be relaxed, calm, and chilled out. Ads accentuate this sense of calm and solitude through an absence of people, both in the vehicle featured and in the background. The technology that figures prominently in these ads frees the driver from the need to manually operate the vehicle and its equipment, or provides a source of amusement, pleasure and relaxation.

In contrast to the theme of the mobile sanctuary, SUVs portrayed as urban assault vehicles emphasize issues of security and defence. Gunster points to the militaristic overtones in ads of this type: "Not only is it a jungle out there, it's also a war" (2004, 20). Escalating apprehensions about crime, violence, and terrorism are reinforced through increasing media attention to these problems, perpetuating the cycle of fear. Campbell claims that these conditions, coupled with a desire for greater personal safety have led to the growth in gated communities and private security guards, as well as the desire for vehicles to be "fortified enclaves (or capsules) secure against others" (2005, 958). He further suggests that the more people encapsulate themselves, the more their fears augment, and in turn, the more they will encapsulate themselves and try to control their personal environment. Urban space becomes something that one has to protect oneself against (Gunster 2004) and civility on the roads gives way to antagonism and hostility. This deflects attention from the public realm and discourages communal activity (Wernick 1991; Wilson 1991).

The resulting culture of aggression and conflict shows up in advertisements through connotations of combat and survival (Schulz 2006). Violence and military overtones become marketing tools: roads and motorways turn into battlegrounds, with other drivers and vehicles as enemies. Each SUV must give its driver a sense of inviolability (Mitchell 2005) and has to be more intimidating and menacing than the next, in order to dominate and control the external environment and those in it. Chrysler's designer, Clotaire Rapaille, has dubbed SUVs "armored cars for the battlefield" and has claimed that "Even going to the supermarket, you have to be ready to fight" (Bradsher 2002, 96–97). Simons (this volume) reinforces this idea through his claim that the structure of urban form was based on military developments and that the modern technologies of the city and automobility characterize what he calls a 'projectile economy'.

Despite ever more threatening external appearances, the contrasting focus on the interior emphasizes refinement and luxury. "The inside of an SUV should be the Ritz-Carlton, with a minibar. I'm going to be on the battlefield a long time, so on the outside I want to be menacing but inside I want to be warm, with food and hot coffee and communications" (Rapaille in Bradsher 2002, 100). Beyond the vehicle's luxury, ads of this type also feature technologies that give drivers complete control of their environment. Bayley asserts that technology is shorthand for "my car is better than yours" (1986, 94), therefore navigational systems (e.g., GPS), power-assisted steering, on-board computers, and mobile communications allow drivers to strategize and win in the 'war' waged against anyone who may intrude or stand in the way of their mobile command centre.

An ad for the Nissan X-TRAIL (*Car* 2004) combines elements of the vehicle as mobile dwelling with militaristic overtones of surveillance and security. The city appears just as dark and hostile as many nature settings but with the added stress of traffic. The caption at the top of the ad recognizes that "It can be hectic sometimes at ground level. Not however, when you're sitting pretty in the Nissan X-TRAIL" and carries on with "Want a taste of the high life?" In the image, the vehicle has not only escaped the nerve-racking traffic, but has transcended it and literally risen above it. By superimposing the X-TRAIL on top of the image, it appears as though it is above the buildings looking down on the streets below. The denoted meaning is that the vehicle provides an escape from the frantic pace of urban traffic comparable to flying above the congestion of the streets. The text highlighting "the 3-D satellite navigation system" and the vehicle's vantage point far above ground level on top of the buildings have deeper connotations of military or security surveillance operations. Other technologies such as 'the smoother and quieter dCi engine' and 'the climate control air-con' allow drivers to be in command of their immediate surroundings while the 'cooling drinks holders' and the sunroof, which permits the occupant(s) to "[watch] the world go by," provide the domestic comforts of the mobile dwelling. The promise that 'you'll be calmed' and 'you can even chill-out' completes the allusion that the occupants are insulated within the vehicle, able to relax and unwind, comfortable and safe, and thereby able to attain a level of freedom even within the road network.

Shifting Meanings: The Anti-SUV Movement and the Next Generation of Ads

Various groups have criticized the expansion of the 4x4 market. Their concerns range from smaller vehicles having their view of traffic ahead blocked and being at greater risk in the event of an accident to problems relating to air pollution and the increased contribution to climate change from higher levels of CO2 emissions. As more people have begun to reconsider the attraction of these vehicles, SUVs have increasingly become contested.

Along with campaigns from traditional environmental groups such as Greenpeace, Friends of the Earth, and the Sierra Club, specific anti-SUV groups have emerged in a number of different cities and countries, as well as on the internet. Examples include the American-based website www.fuh2.com, with its message 'F.U. and your H2', which allows people to submit photos of themselves giving 'the official H2 salute' to Hummers. Also in the US, the Evangelical Environmental Network had a campaign with the slogan "What would Jesus drive?"

Campaigns by the UK's Alliance Against Urban 4x4s have included placing 'parking tickets' on SUV windshields, offering 'mud washes' so that drivers can give their vehicles a 'real off-road look', and lobbying for increased congestion charging and higher road taxes for SUVs. An action held by the group on Valentine's Day (Alliance Against Urban 4x4s 2007) took a more positive spin on the anti-4x4 campaign: rather than demonizing SUVs, they rewarded drivers of small fuel-efficient cars by leaving Valentine's cards on windshields, thanking the vehicle owners for making an environmentally-conscious choice, and for caring about the future. Other groups have taken a more overtly political stance. The Detroit Project, for example, has run ads in the US that associate 4x4s with the funding of international terrorism through over-reliance on Middle Eastern oil. Their goal has been to show the connection between individual choice and political effect by linking improved fuel efficiency to national security (Campbell 2005).

The rebranding of Jeep, Land Rover, and Toyota Land Cruiser into "JEEPollue" ['Jeep Pollutes'], "Land Pollueur" ['Land Polluter'] and "Land Cuiseur" ['Land Cooker']⁴ by the French group Anti4x4 is an example of the subversive tactics of culture jamming, which provide what Bordwell describes as "perspective by incongruity" (2002, 247). The appropriation of recognized brand identities or logos in order to redefine a product with an alternative meaning, which Heffner et al. (2006) call bricolage, is intended to alert people to the dangers of a consumption-based society encouraged by the advertising industry and a capitalist economy. The result is a 'meme warfare', with battles waged over symbols of representation and the meanings that shape ideologies and values in order to shift the dominant paradigm (Bordwell 2002; Carducci 2006).

Oppositional or resistant readings, counter-ideologies and hostile criticism provide new forms of appropriation and new messages for advertisers to incorporate into the next generation of ads (Carducci 2006; Dyer 1982). For example, Linder (2006) points to the success of social and environmental marketing campaigns whose meanings have subsequently been appropriated by advertisers looking to enhance their product's social standing. Johnson's (1986) model of the circuit of culture, further developed by Carvalho and Burgess (2005), identifies three interdependent phases which shape the meanings of media and cultural objects: production, circulation, and consumption. Using the case of SUV advertising as an example, producers encode advertisements with messages or particular meanings which they then disseminate to target audiences. Next, the viewers go through a process of decoding the meanings; however, because viewer decodings are based on the context of their social relations and lived cultures, the advertisers' message may not be read as intended (for a similar analysis of popular culture, see Genovese, this volume). As Carvalho and Burgess (2005) explain in their adaptation of Johnson's model, when viewers read messages in ways other than originally intended, or when attitudes or contexts change, the circuit shifts with the new contexts providing the raw material for the next stage of production. Therefore, a new generation of ads appropriates and recirculates the social and environmental arguments directed against SUVs in order to rebuff criticism and ease people's anxieties and channel their fears into consumption-based desire (Dyer 1982; Linder 2006).

Gunster claims that the objections lodged against SUVs have reinforced "the logic that has helped to establish cross-over vehicles" (2004, 15). People who dislike large SUVs can have a smaller four-wheel drive vehicle instead, believing that they are doing less harm to the environment, which is precisely how advertisers portray them. On the other hand, ads depict full-size 4x4s "as 'genuine' off-road vehicles" (15). Another way that advertising dispels opposition is to incorporate it into their ads (Dyer 1982), often through the use of humour. For example, in response to the criticism that very few drivers actually take four-wheel-drive vehicles off-road, a 2004 ad for the Honda CR-V had the tagline "Wilderness tbc"

---

4   See Anti4x4 website.

and an ad in Fall 2006 showed an SUV on a residential street approaching a large pothole with a comment that the driver actually does go 'off-road'. As another way of addressing this concern and acknowledging that SUVs are now mainly driven in cities and on highways, there appears to be a trend towards producing a greater number of ads with urban settings or using studio backdrops, particularly in the case of crossover vehicles.

In dealing with concerns about environmental damage, advertisers have used different approaches. Marshall (2001) suggests that people respond to environmental concerns in one of three ways: they completely ignore them; they acknowledge them but totally defy them; or they acknowledge them and show how they are addressing them. Linder (2006) adds a fourth response: people may adopt a sceptical stance regarding claims of any adverse effects. Although individuals may accept that climate change is in fact occurring, they may do so without actually accepting the responsibility or the moral implications of the consequences. For example, many people believe that it is up to the government or industry to make changes and they overlook their own role in the process. Cohen calls this the "passive bystander effect" where individuals "wait for someone else to act and subsume their personal responsibility in the collective responsibility of the group" (cited in Marshall 2001, N.p.). When a person is in denial or has taken on a sceptical position, this can take the form of intentionally partaking in wasteful or environmentally-unfriendly activities and practices and blatantly disregarding their impacts.

Although Marshall's (2001) paper examined the process of denial in terms of the behaviour of the general public, indulgence in deliberately wasteful activities and disregard for the results can also apply to industry. Jeep does just this in a Cherokee ad (*What Car?* 2005). In accepting the criticisms against SUVs, Jeep mockingly admits to being 'The Original Dirty Four Letter Word'. The image shows the Jeep skidding on a dirt road, trailing a cloud of dust behind it (and because the background is blurred, we can assume that the vehicle is actually being driven in this way), giving a second meaning to the obvious connotation of 'original dirty four letter word'. As one of the first 4x4s to be built, the meaning of 'original' is clear, but the ad also appropriates the notion of genuine off-road vehicle.

Another Jeep ad, this time for the Grand Cherokee, shows the vehicle sitting on top of a calm, flat body of water with the caption 'The end of the world is never nigh'. The denial inherent in both these ads and "the dismissal of the consequences of one's actions, or at least their dismissal in the face of the more highly valued victory in the game of driving, encapsulates both the theme of the individualist and the overarching 'rights' to go anywhere" (Paterson and Dalby 2006, 19). These ads provide an ironic reversal of the message of restraint and self-discipline inherent in much environmental advertising and offer an escape from any associated anxieties or guilt. By acknowledging the negativity associated with their product, the advertisers aim to banish the consumer's fears and neutralize any objections (Wernick 1991). However, Linder asserts that this "covert assault on risk messages" (2006, 131) devalues and diverts public concern and cheapens the symbols and meanings used in social and environmental campaigns.

Manufacturers such as Lexus have taken a different approach in a series of advertisements for the RX400h (*The Economist* 2005). In these ads, they are not only acknowledging environmental concerns and the criticisms against gas-guzzling SUVs, but drawing on the theme of accountability to build a reputation as a socially and environmentally-responsible company (Linder 2006). As the first hybrid SUV, one of the ads focuses on the 'breakthrough in engineering' made possible by scientific advances. "The purpose of such technologies is to sustain consumer needs; the need for comfort, control, power, and, at least pejoratively, for projecting an image of environmental consciousness" (Meister 1997, 230). Unlike the other ads discussed above, this one is on a plain backdrop, with both the background and the vehicle shaded in tones of green. The tagline "World changing not planet changing" further positions this SUV as a 'green' or environmentally-benign product and aims to diminish the drawbacks of automobility (Paterson and Dalby 2006). The tactics of green advertising are particularly persuasive because they resonate with public concern for the environment (Hansen 2002; Holder, 1991). However they are also inclined to be misleading and deceptive. For example, although the RX400h may be a hybrid, its fuel consumption figures and CO_2 emissions ratings are actually worse than some fuel-powered SUV models.[5] Two other ads in the series show X-ray images of the RX400h's internal mechanical equipment superimposed over the vehicle body. Williamson (1978) notes that this photographic technique gives the impression of full disclosure, transparency and having nothing to hide. Once again environmental messages are appropriated, but this time to enhance the product's social standing. The goal is to allay people's fears, ease their consciences and dispel any guilt that may discourage continued consumption.

While many factors undoubtedly contributed to the rise in the SUV's popularity, this chapter suggests that advertising played a key role in the expansion of the 4x4 market by designing images and constructing symbolic meanings which powerfully tapped into deeply held ideological beliefs, shaped buying behaviour and influenced consumption-related practices. With increasing concerns about climate change, growing criticism directed towards SUVs and their manufacturers, a future of

---

5   According to figures listed in the advertisements I analyzed, the RX400h has the following fuel consumption figures in Ltr/100km (mpg): 9.1 (31.0 mpg) Urban; 7.6 (37.2 mpg) Extra-urban; 8.1 (34.9 mpg) Combined; 192 g/km CO2emissions. The Nissan X-TRAIL performs better in each category with figures of 9.0 (31.4 mpg) Urban; 6.2 (45.6 mpg) Extra-urban; 7.2 (39.2 mpg) Combined; and 190 g/km CO2. The Honda CR-V Diesel beats them both with 8.2 (34.4 mpg) Urban; 5.8 (48.7 mpg) Extra-urban; 6.7 (42.2 mpg) Combined; and 177 g/km CO2. The Land Rover Freelander gets better fuel consumption at the top end of its ranges in each of the categories, and even the Jeep Cherokee and VW Touareg do better on Extra-urban driving at 7.5 L/100km (37.7 mpg) and 5.9 – 7.8 L/100km (36.2 – 47.8 mpg) respectively.

higher gas prices, and an economic crisis that has hit the auto industry, we can expect shifts in advertising meanings over the coming years: in representations of nature and civilization, in the selective reintegration of counter-cultural messages into commercial discourses, and in the development of new techniques of branding and promotion to catch the attention of increasingly jaded consumers.

Appendix

Table 3.1 below lists the magazines that were used for finding SUV advertisements to analyze. All monthly issues (from January 2004 to September 2005) of the motoring, men's and women's lifestyle, and travel magazines were reviewed, while a number of issues of each of the magazines in the miscellaneous category were examined. The predominant number of ads came from the motoring and men's lifestyle magazines, with most of the same ads being run in each of the magazines. *Wanderlust* and *Geographical* had no SUV ads until early 2005, and then only one ad per issue, usually the same ad in each, and almost always for the Nissan X-TRAIL. The women's magazine, *Zest*, had no SUV ads at all and, in fact, had very little car advertising. *SHE* magazine had an SUV ad in seven of their issues in 2004, but only one in the period of January to September 2005. It is interesting to note that only the ad run in 2005 had a nature setting. SUV advertising with a nature setting was prominent in the motoring and men's magazines, thereby implying that adventure in the outdoors and the mastery of nature is a domain that appeals to men more than women, or at least from the point of view of the advertisers.

Table 3.1 List of Magazines Used in Advertisement Search and Analysis

Motoring	Men's Lifestyle	Women's Lifestyle	Travel	Miscellaneous
Car	*Men's Fitness*	*SHE*	*Condé Nast Traveller*	*Business Week*
EVO	*Men's Health*	*Zest*		*The Economist*
What Car?				*Geographical*
4 x 4				*Wanderlust*

References

Advertise CO2. http://www.advertiseco2.eu (accessed 23 June 2008).
Alliance Against Urban 4x4s. 2007. Valentine's Card. http://www.stopurban4x4s. org.uk (accessed 13 February 2007).
Anti4x4. http://www.anti4x4.net (accessed 11 February 2007).

Bayley, Stephen. 1986. *Sex, drink and fast cars: The creation and consumption of images*. London: Faber and Faber.

Bordwell, Marilyn. 2002. Jamming culture: Adbusters' hip media campaign against consumerism. In *Confronting consumption*, Thomas Princen, Michael Maniates, and Ken Conca, eds, pp. 237–53. Cambridge: MIT Press.

Bradsher, Keith. 2002. *High and mighty: SUVs – The world's most dangerous vehicles and how they got that way*. New York: PublicAffairs.

Campbell, David. 2005. The biopolitics of security: Oil, empire, and the sports utility vehicle. *American Quarterly* 57(3): 943–72.

Car. 2004. Nissan X-TRAIL. September: 109.

Car. 2005. Volkswagen Touareg. June: 23.

Carducci, Vince. 2006. Culture jamming: A sociological perspective. *Journal of Consumer Culture* 6(1): 116–38.

Carvalho, Anabela and Burgess, Jacquelin. 2005. Cultural circuits of climate change in UK Broadsheet Newspapers, 1985–2003. *Risk Analysis* 25(6): 1457–69.

Craig, Robert. L. 1992. Advertising as visual communication. *Communication* 13(3): 165–79.

DuPont. 2006. Global automotive color: 2006 popularity. *2007–2010 Trends*. Troy, Michigan: DuPont Automotive Systems.

Dyer, Gillian. 1982. *Advertising as communication*. London: Methuen and Co.

Garnar, Andrew. 2000. Portable civilizations and urban assault vehicles. *Techné: Journal of the Society for Philosophy and Technology* 5(2): 1–12.

Geraghty, Christine. 2000. Representation and popular culture: Semiotics and the construction of meaning. In *Mass Media and Society*, James Curran and Michael Gurevitch, eds, pp. 362–75. London: Arnold.

Gunster, Shane. 2004. 'You belong outside': Advertising, nature, and the SUV. *Ethics and the Environment* 9(2): 4–32.

Hansen, Anders. 2002. Discourses of nature in advertising. *Communications* 27(4): 499–511.

Heffner, Reid R., Turrentine, Thomas S. and Kurani, Kenneth S. 2006. A primer on automobile semiotics. University of California, Institute of Transportation Studies: Davis, CA. Research Report UCD-ITS-RR-06-01. http://repositories. cdlib.org/cgi/viewcontent.cgi?article=1106&context=itsdavis (accessed 7 May 2007).

Holder, Jane. 1991. Regulating green advertising in the motor car industry. *Journal of Law and Society* 18(3): 323–46.

IOL. 2007. Colour fast. *Independent Online* 8 May. http://www.iol.co.za/general/newsview.php?click_id=661&art_id=iol1178624374146C461&set_id=16 (accessed 31 May 2007).

Jain, Sarah S. Lochlann. 2002. Urban errands: The means of mobility. *Journal of Consumer Culture* 2(3): 385–404.

——. 2005. Violent submission: Gendered automobility. *Cultural Critique* 61(fall): 186–214.

Johnson, Richard. 1986. The story so far: And further transformations? In *Introduction to contemporary cultural studies*, David Punter, ed., pp. 277–313. London: Longman.

Kress, Gunther and van Leeuwen, Theo. 2006. *Reading images: The grammar of visual design*. Abingdon: Routledge.

Linder, Stephen H. 2006. Cashing-in on risk claims: On the for-profit inversion of signifiers for 'global warming'. *Social Semiotics* 16(1): 103–32.

Macfarlane, Robert. 2005. 4x4s are killing my planet. *The Guardian* 4 June: 34.

Marshall, George. 2001. The psychology of denial. http://ecoglobe.ch/motivation/e/clim2922.htm (accessed 15 July 2008).

McLean, Fiona. 2005. SUV Advertising: How visual images in the media promote unsustainable consumption. Unpublished MSc Dissertation. London: Department of Geography, University College London.

McQuarrie, Edward F. and Mick, David Glen. 1999. Visual rhetoric in advertising: Text-interpretive, experimental, and reader-response analyses. *Journal of Consumer Research* 26(1): 37–54.

Meister, Mark. 1997. 'Sustainable development' in visual imagery: Rhetorical function in the Jeep Cherokee. *Communication Quarterly* 45(3): 223–34.

Men's Health. 2004. Volkswagen Touareg. October: 8–9.

Men's Health. 2005. Land Rover Discovery. January/February: 6–7.

Messaris, Paul. 1992. Visual 'manipulation': Visual means of affecting responses to images. *Communication* 13(3): 181–95.

———. 1997. *Visual persuasion: The role of images in advertising*. Thousand Oaks, CA: Sage.

Mitchell, Don. 2005. The SUV model of citizenship: Floating bubbles, buffer zones, and the rise of the 'purely atomic' individual. *Political Geography* 24(1): 77–100.

Paterson, Matthew. 2007. *Automobile politics: Ecology and cultural political economy*. Cambridge: Cambridge University Press.

Paterson, Matthew and Dalby, Simon. 2006. Empire's ecological tyreprints. *Environmental Politics* 15(1): 1–22.

Rose, Gillian. 2001. *Visual methodologies: An introduction to the interpretation of visual materials*. London: Sage.

Schulz, Jeremy. 2006. Vehicle of the self: The social and cultural work of the H2 Hummer. *Journal of Consumer Culture* 6(1): 57–86.

Scott, Walter Dill. 1908. *The psychology of advertising: A simple exposition of the principles of psychology in their relation to successful advertising.* Cambridge, MA: Small, Maynard and Co.

Sheller, Mimi. 2004. Automotive emotions: Feeling the car. *Theory, Culture and Society* 21(4/5): 221–42.

Stokes, Gordon and Hallett, Sharon. 1992. The role of advertising and the car. *Transport Reviews* 12(2): 171–83.

The Economist. 2005. Lexus RX400h. 9 July: 28–29.

Urry, John. 1999. Mobile cultures. Department of Sociology, Lancaster University. http://www.lancs.ac.uk/fss/sociology/papers/urry-mobile-cultures.pdf (accessed 20 April 2005).
——. 2000. Inhabiting the car. Department of Sociology, Lancaster University. http://www.lancs.ac.uk/fss/sociology/papers/urry-inhabiting-the-car.pdf (accessed 20 April 2005).
Wernick, Andrew. 1991. *Promotional culture: Advertising, ideology and symbolic expression.* London: Sage.
What Car? 2005. Jeep Cherokee. April: 126–27.
Whitely, Nigel. 1999. Readers of the lost art: Visuality and particularity in art criticism. In *Interpreting visual culture: Explorations in the hermeneutics of the visual*, Ian Heywood and Barry Sandywell, eds, 99–122. London: Routledge.
Williams, Raymond. 1982. *The sociology of culture.* New York: Schocken Books.
Williamson, Judith. 1978. *Decoding advertisements: Ideology and meaning in advertising.* London: Marion Boyars Publishers.
Wilson, Alexander. 1991. *The culture of nature: North American landscape from Disney to the Exxon Valdez.* Toronto: Between the Lines.

Chapter 4

Bad Impressions: The Will to Concrete and the Projectile Economy of Cities

Derek Simons

Figure 4.1 Hemlock Street Overpass, Vancouver, BC. In the early hours of 17 August 2007, a truck lost control and catapulted over the railing of this overpass in Vancouver. It landed upside down on the street below and exploded, killing the driver, and hospitalizing a passenger with critical injuries.

Concrete has a long history, yet its richly-imbued cultural, social and political signification in automobility's built environment is largely unacknowledged. As a fundamentally mouldable material that underlies most infrastructures, concrete stretches back as far as the ancient Roman Pantheon and forward to modern automobility's roads. This mundane material joined other technologies transformed during the Great War – most notably, the locomotive, the explosive military projectile and the camera – into a projectile economy that created a radically new culture of space and time, based on speed. In tracing these historical antecedents of modern mobility and considering their cultural and psychic meanings in relation to the self, this material investigation of culture is necessarily interdisciplinary and conjectural: it illustrates linkages between the material, the semiotic and the self. It argues that such technologies of impression as concrete – technologies centred on a force or action impressed upon or moulding a receptive surface – have acquired a power and energy central to modernity's thrilling projectile economy and complementary technologies of the self that connect truth and sacrifice. Impressive technologies, whose origins are out of view, have become pervasive in modernity, on the landscape, the built environment, human bodies and the self. As constituents of the automobility system, they help to explain what sustains the automobile as a dominant mode of transportation, despite its many troubles. In exploring the modern self's yearnings for the thrill of the automobile, this study uncovers reasons for its willingness to accept the costs of routine morbidity inherent in the dialectic of speed and collision.

According to the Insurance Corporation of British Columbia, 29,625 people were killed or injured by automobiles in the province in 2004 (ICBC 2004). The number of dead quickly becomes very large; between 2000 and 2005 in Canada as a whole, for example, 17,053 people died on its roads (Transport Canada 2005). Injury should not be minimized either. Most Westerners can relate stories of how the lives of relatives and friends or of themselves have been fundamentally altered by automobile injury.

These casualties are only the direct consequence of the automobile; its enormous additional indirect environmental and social cost is well-known. On balance then, we can easily see that the automobile guarantees a destructive legacy. Can we in light of this egregious guarantee speak then logically of 'car accidents'? To embrace so happily the allures of the wheel and road, minds must necessarily be shut to the hospital visits, rehabilitative therapy and funerals that are automobility's inescapable dark side.

If the notion of accident is discarded, though, how then might we understand the apparent willingness to endure such certain costs? While the powers of advertisers and car stylists are strong, are they sufficient to explain the acceptance of the litany of injury and death recited so nonchalantly in radio traffic reports every morning and afternoon?

Jeffrey Schnapp (1999) in a wonderfully suggestive article explores speed as a defining if under-acknowledged agent of modernity (which Genovese evocatively contextualizes in her study of drag racing, this volume). Schnapp argues that the sheer brute pleasure of going fast culminating in the automobile has driven technical innovation since at least the 1600s. But his account still leaves us wondering. Is speed a 'drug', as Schnapp describes it, unleashed by modernity, a long-latent but murderously dangerous propensity to feel the wind on our faces and the landscape rushing past? Is this latent propensity the reason why car advertisers so keenly highlight this form of excitement, as Conley shows (this volume)?

In this chapter I argue that the answer is both yes and no. No, speed in and of itself cannot suffice as the missing causal agent of automobility. But yes, powerful and deep yearnings related to speed are so old as to be effectively if not ontologically latent. These yearnings of the modern self might be characterized as aching for something very old: sacrifice. Sacrifice is often used unhesitatingly to describe loss of life in warfare (Doyle 2007),[1] but the word's implicit wilfulness seemingly inspires reluctance to characterize similarly the routine morbidity of automobility. But Schnapp goes further in arguing that a will-to-death is the central characteristic of the dialectic of speed and collision inherent to the vast accumulation of people (passengers, drivers) and things (freight, commodities) brought together at high rates of velocity that Schivelbusch in sum calls the 'projectile economy'.

In this chapter, I want to get at this sacrificial will-to-automobility within the projectile economy by considering the silences about and turnings away from the built infrastructure. This mostly urban infrastructure is the rarely acknowledged but necessary constituent of automotive existence and human injury and death. These silences set the stage for a critical reconsideration of the role of the built environment in mobility, and more specifically automobility. By studying some historical antecedents of the widespread infrastructure of mobility – especially the concrete barrels through which the missiles of the automotive economy hurtle – the ontological stakes raised by the river of the automobile's damaged bodies become more apparent.

My approach to studying the deadly infrastructure of automobility will marry art historical analysis of infrastructure as visual culture to insights from science and technology studies that challenge boundaries between the material/objective and the social/subjective (Latour 1992; De Landa 1992), culminating in a Foucauldian conclusion of the centrality of sacrifice in the Western development of truth through the self conceived as a technology. It is here, I argue, that the necessary correlation between death and automobility can be found.

To begin then with silences, consider reactions arising from encountering road accidents. People feel empathy for the victims of course, but also an urge to see what has happened, to drink in the full scale of the horror, perhaps even to wish subconsciously for the mayhem to have been more dreadful. J.G. Ballard's novel *Crash* (1973), later made into a film by David Cronenberg (1996), implicates readers and viewers in a psychosexual struggle between sex and death, Eros and

---

1   In marking recently Canada's 90th anniversary of the Battle of Vimy Ridge, TV reporting talked loftily of the notion of blood sacrifice as necessary to define nationhood. As Doyle (2007) noted, nobody on TV seemed to question this idea's validity.

Thanatos, that is inherent to car crashes. (Schivelbusch similarly relates amazed early nineteenth century commentaries that claimed male victims of spectacular train crashes had died with erections.) The will-to-see and the will-to-not-see car crashes engage in a complicated dance that is uncomfortably close to pornographic. But an even deeper obviation takes place: the role of built form. Even when the built environment has a central role in accidents as proximate cause, as it often does, we usually relegate its functioning to the unimaginative world of engineering and risk management.

And this carries over to critical studies of automobility. Most critical studies of automobile collisions overlook infrastructure. Dennis Soron (this volume), for example, acknowledges material environments among the social structures in which everyday lives are embedded, but he concludes with a call for a post-consumerist politics, in the course of which built structures will presumably take care of themselves. Similarly, Fiona McLean's chapter on SUV advertisements joins other work on mediated automobility and such contiguous realms as marketing, risk, alcohol drinking or race relations. I am not contesting the importance of such work; on the contrary. I am however highlighting how little critical curiosity is piqued by what exactly goes on in the material realms of concrete and asphalt.

Somewhat solitarily then, I focus here on concrete, the most fundamental material component of most infrastructure, and in so doing turn the usual formulations of the relation of the causal social and cultural and the merely resultant built environment on their head; my question is, what peculiar energies move within or through concrete that have created the social questions to which concrete then appears as the inevitable built answer? To explore this question, I examine technologies of impression – those that involve making a direct mark upon a receptive or mouldable surface – that radically reconfigured the world: from Roman uses of such powerful impressions as concrete buildings and masks to more recent transformations of mobility, militarism, images and the built environment, particularly during the Great War. This discussion illustrates how mundane materiality such as concrete acquires a power that has become central to modernity's thrilling projectile economy and complementary technologies of the self that connect truth and sacrifice.

A determinedly material investigation can begin by consideration of the forms of concrete. What general characteristics can we ascribe to them? Concrete is uniquely hard, and it is cheap, for example. But the quality of concrete that trumps all others is its mouldability. Concrete is moulded to accommodate a specific need or needs. The overpass or sidewalk will not easily brook any use other than its original intention. Concrete is unique among building materials in this respect. It is identical with the structure it creates. It cannot be removed from a structure without destroying it. In short, its form is an impression of the activity for which it is created.

Concrete's Impressive Origins

The Oxford English Dictionary tells us the Latin origins of the word 'impression' were military, referring to irruption, onset and attack. Impression can be linked etymologically to missile, an object thrown over space. The military and social implications of these words, missile and impression, seem over time to have transferred. Missile originally had some peaceful connotations, referring to the offerings thrown to a crowd by a ruler, a sense that lives on in the related word, missive. Impression meanwhile took on its more civilian meanings of stamp or powerful excitation only with Cicero. The conflation of these various meanings of missile and impression – moulding or stamping, ability to excite feelings of awe, transmission by a ruler over space, and violent surprise or attack – are evident in the dawn of concrete architecture, a century or so after Cicero, especially in emperor Hadrian's most famous concrete structure, the Roman Pantheon. Stendahl joined a long line of enthusiastic commentators when he described this building as the "finest remnant of antiquity's ... sublime ... daring" (Meeks 1960, 157). The Pantheon is, in a word, impressive, the early predecessor of modern structures from Speer's Volkshalle to the Houston Astrodome designed to awe and engulf spectators; directly connecting the minutely profaned individual and the infinitely transcendent (whether conceived as spirit, politics or entertainment) through voluminous space at its most sheer.

Interpretation of the Pantheon's intended original function is necessarily speculative, and the building has been variously considered: as a monument to Saturn or to the gens Julia and divine ancestors; as a solar temple; as a manifestation of ideal geometry or Pythagorean mathematics; or as a temple to imperial Rome and all things Roman in the world (Joost-Gaugier 1998; MacDonald 1976; Wilkins 2004). Hadrian was absent from Rome during much of the time during which the Pantheon was constructed (in part no doubt because both the Senate and the Roman people viewed him with some ambivalence and even hostility). Thus we can add to the speculative list of architectural purposes of the Pantheon that it served to solidify the new emperor's political and spiritual position, despite the absence of the emperor's person. Surely for an emperor who proved to be unusually effective, and for a people as acutely conscious of the rhetorical function of imperial symbolism as were the Romans, tangible manifestation of the transcendent power of empty space, standing in for the absent emperor, must have had considerable importance.

Hadrian's builders could accomplish this magnificent spatial sleight-of-hand because of their skill in hydraulic concrete, at the time a relatively newly discovered building material. The Pantheon was not only impressive, but was itself impressed, formed by careful moulding of concrete over the 142 feet of its dome's diameter (Winter 1979, 137). Thus we have an uncanny confluence of technical ability and propagandistic purpose, in which an impressive material capable of conveying forcefully the power of absence served well the specific imperial politics of the High Roman Empire.

The Pantheon reveals a complex mixture of absencing and presencing work. In the first instance the structure self-evidently revolves around a void. Its architectural condensation of absence takes on a visceral reality or presence, standing in rhetorically for Hadrian as the absent emperor. To have been physically present would perhaps have lost him the imperial title (there were attempts on his life when he first returned to Rome as emperor), but to have been wholly absent would have had the same result. The Pantheon may have circled the square of this dilemma. Later, Hadrian seems to have used the temple for civic functions, dispensing justice from a perch in the apse (Joost-Gaugier 1998), perhaps thereby fulfilling for Romans the prophecy of his power made by the Pantheonic dome's void. But as a student of Stoicism[2] he was also acutely aware of the limits of his life and imperial reach, and just as the space of his temple may have foretold of the power of his presence when he was still absent, so too may it have foretold of his ultimate absence to come through death, even as he acted out the power of his presence through his ultimate judicial authority as emperor. As his absence belied his power, which the Pantheon unarguably reaffirmed, so too did the space of the temple remind that his powerful, justice-dispensing presence belied his weakness, his mortality, in which he was, as Marcus Aurelius (1992) poignantly described, just like everyone else. Thus the Pantheon physically stands in for ambivalence around the identity and power of the emperor: present when absent, absent when present.

None of Hadrian's semiotic efforts in the Pantheon would have made sense if the Roman people were not equipped to interpret them, and indeed absencing and presencing was important in daily Roman life (at least of the patrician and noble classes, doubtlessly for Hadrian the only ones that mattered) in the funerary traditions from the republican era. Making wax casts of the faces of the dead, depicted in the famous statue Patrician Carrying Busts, was a very old Roman tradition. By some accounts these cast masks would be worn by actors in front of the bier of the deceased on their way to burial (Goldscheider 1945).[3] The Latin word persona evolved from these masks, which self-evidently performed on a micro scale the absencing and presencing work performed more grandly by the Pantheon. The moral verity of the absent ancestor was guaranteed by the individual's unique impress of worry lines and scars upon the mask, a guarantee that many assert underlay the emergence of the remarkable Roman busts known as veristic sculpture.

---

2   Stoic philosophy seems to have had a strong influence at the centre of the emperor's judicious handling of the various unruly elements of empire. Central to Stoic philosophy was a theory of impression. The good, the true and the just all depended on acceptance of things as they impressed themselves upon human consciousness.

3   After the funeral, the masks of the ancestors, known as imagines, including the most recent addition to the ranks of the dead, would be stored in special cabinets or shrines called imagines majorum in the family homes, awaiting their display at the next significant public event (Swift 1923).

In Vitruvius's famous books on architecture (Vitruvius 1914), this guarantee is evident in his chapter on concrete and other building materials. He goes out of his way to describe himself as ugly and old, evidently to buttress the validity of his claims about building. As Greeks projected a classically handsome sense of themselves on their shields, so too did Romans such as Vitruvius sketch in words or in sculpture their own self-consciously flawed faces (Flower 1997). It is further made apparent that Vitruvius revels in handling concrete as work perfectly appropriate in its rough qualities for the flinty republicanism apparent in veristic sculptures and his own self-description.

From almost its first moments of even its more pedestrian uses, then, concrete was imbued with signification, reflecting back the tough and powerful self-appraisals of a class acutely conscious of the semiotic power of impression as exemplified by the physiognomy of ancestor death masks stored in all their homes. It can thus be imagined more easily how Hadrian's designers would have happily pushed concrete's semiotic boundaries toward the Pantheonic apotheosis of space.

One final note on antique technologies of impression, thanks to Joseph Rykwert's (1976) seminal work, we know that the tools the Pantheon's designers used were richly imbued with spiritual meanings, hearkening back to magically resonant foundational Roman practices in which specific spaces – especially intersections of streets – were anointed with sacrifice. Rykwert focuses in particular on augurs, Roman soothsayers who sought signs indicating the fate of particular spaces. The measuring and plotting tools used by augurs were only secondarily about creating and transferring measurements and more centrally concerned with the relays between the specific and the divine indicated by various forms of divination, including sacrifice and reading entrails. We can assume then that, to the Romans who used them, other related tools such as callipers, dividers, set squares and rulers resonated beyond the purely functional qualities apparent to modern sensibilities. Moreover, these tools from their earliest uses were closely linked in practice to etching in various impressionable materials such as wax, vellum or parchment.

We can assume some of this magical relationship between measuring tool and the divine – in which divinity was impressed in both site and representation – would have clung to the instruments used by builders in the Dark Ages, even though the capability to make precision scoring tools was lost. As architecture and building re-emerged into the textual era of the Renaissance, the importance of using building tools as a means of impressing representations of space on more portable surfaces reasserted itself, and endured even after the introduction of paper that replaced the more impressionable vellum, wax and papyri surfaces that survived the classical collapse. Scored drawings, sometimes later filled in with ink, in fact continued until the Renaissance and even survived in use until the 1700s (Hambly 1988), when precision drawing tools became wholly centred on the pen riding over the surface of paper.

Impressive Technologies, the Great War and the Modern City

At the same time as the switch to paper drawings, impression elsewhere reasserted itself as a central motif, and especially through four drastically different ways of making impressions that radically reconfigured the world: the locomotive, the explosive military projectile, the camera and concrete. Each of these technologies at their core relied upon transmissions making a direct mark upon a surface. The locomotive's track requires a landscape and human sensorium moulded to reflect the train's progress. The explosive projectile impresses power upon the bodies and walls of those who would resist. The camera impresses light emitting from an object upon paper. Concrete impresses actions on space. Without these impressive technologies, we can safely conjecture, modernity would have been radically different, if it would have appeared at all. Each of these four technologies had a long history but emerged at the end of the medieval era in a still-primitive form to which we may imagine still clung their ancient magical origins; each underwent significant transformation in the early to mid-1800s; each was probably most influential in the social, cultural and mechanical confluence of the city; and all morbidly combined, transformed and accelerated in World War I, setting the catastrophic boundaries of the century to come.

Woven together (Crary 1999), these technologies formed, as we have already noted, what Wolfgang Schivelbusch (1986) has memorably characterized as the projectile economy. Schivelbusch came up with the phrase, the projectile economy, to describe the world-making effects of the railroad. Schivelbusch picks up the story of the railroad where the ancient, slow and localized development of the wagonway was dramatically transformed by the rapid introduction of newly invented engines in roughly a forty year span at the very end of the eighteenth century, with far-reaching consequences. Schivelbusch suggests the myriad psychophysical changes that responded to the challenges posed by the railroad were defence mechanisms working on multiply minute and macro-scales: new spring technology; new psychic mechanisms (he uses the Simmelian and Freudian terminology of 'stimulus shield') to cope with the strains of train travel, including a large increase in reading among bourgeois travellers; panoramic vision, wherein the foreground appears as a blur while the farther distance slips by at a statelier rate; remaking of the countryside and the cityscape; implementation of standard time. Taken as a whole, the railroad engendered nothing less than a new culture of space and time (Kern 1983), as its mobile successor the automobile continues to do today (see Dennis and Urry's chapter).

Shivelbusch (1986) gives a couple of examples of contemporary reporting that likened mid-nineteenth century trains (then capable of speeds of roughly 75 miles per hour) to a specific kind of projectile: the cannonball. It is outside of Shivelbusch's purview that cannonballs themselves were undergoing a similar recent metamorphosis.[4] The development at the beginning of the nineteenth century of milling machines capable of cutting component parts into prescribed shapes transformed armament production and eventually, especially after the Crimean War later in the century, the forms of conflict.[5] Although these first efforts were of dubious actual efficiency, the principle of machinic interchangeability spread to other theatres of war, especially after many influential British were impressed by Samuel Colt's display of revolvers with interchangeable parts at the Great Exhibition of 1851 in London, very close in time and place to the Britons who, Schivelbusch relates, compared the experience of train travel to being shot out of a cannon.

The structure of urban form both reflected and drove these new military developments and the projectile economy was about to engender another transformation of the impression/transmission dialectic. The cannonballs that could be fired in 1850 merely doubled the performance of those in 1600;[6] 30 kilogram balls could still be hurled at most only one and a half kilometres. By 1880, though, cannons could shoot 100 kilo explosive shells over 6 kilometres, and by 1914 the Germans could fire shells weighing a tonne. Thus many fortresses became pointless, since these shells could crack any but the most advanced reinforced concrete structures, such as the fortress at Verdun; many poor Belgian defenders huddled inside forts in 1914 went insane listening to the systematic explosive advance of the invading Germans' massive shells destroying their obsolete structures (Hirst 2005).

The suddenly obsolete fortress was superseded by the easily built and replaced trench, since piled dirt could render even large calibre shells only very locally effective. Over the First World War trenches became the organizing principle of combat. The Germans developed the feste, or fortified, group principle (Hirst 2005) in which they distributed mutually supporting concrete bunkers irregularly over the terrain of the battle front. This was highly effective, but the Germans were eventually overwhelmed by the sheer numerical superiority of their Allied adversaries.

Hirst (2005) notes that the massive concrete and mechanized fortresses predicted modernist cities. It is a prophecy worth spelling out in more detail. Like the modern cities to come, the fortresses of World War I both relied upon and had

---

4   The mechanically powered projectile, such as the catapult or the crossbow, had a long history, but they were overtaken by a new transmissive agent: gun powder. After a long, slow gestation, gun design had remained essentially static for almost half a millennium (McNeill 1982). Lighter, more powerful, and accurate guns emerged late in the 1400s.

5   These innovations first appeared at the Springfield Massachusetts armoury, where production operated according to the principle of interchangeability of parts (Hounshell 1984).

6   As production techniques improved in the seventeenth century, a master engineer such as Vauban in France was able to say: "There are no more just judges than cannons, for they go straight to the point and cannot be corrupted" (Virilio 1994a, 19).

to solve the problem of a massive influx of a hitherto primarily rural population into centralized spaces of battle. This hoovering up of whole populations was made possible by railroads. The French for example mobilized nearly three million men in August 1914 using 4,278 trains (Kern 1983). To accommodate these numbers, fortresses from the 1880s on featured such urban innovations as retractable armoured cupolas, mechanized ventilation systems and elevators.

Only concrete was capable of generating structures with enough speed and intrinsic strength to accommodate the specific needs of this huge influx of armed men. Concrete had been developed continuously in France and the United Kingdom from the end of the eighteenth century, but like artillery and railroads, it had transformed in the middle of the nineteenth century as its components were systematically analyzed. Its use remained limited however. Prior to World War I, concrete's value was resisted out of aesthetic, engineering and safety concerns, but the emergency conditions of war transformed its semiotic powers from repulsive to requisite, and thereafter the use of concrete became widespread (Collins 2004).

Initially concrete appeared to be a replacement for masonry (we recall here McLuhan's 1997 aphorism that progress drives into the future looking in the rear-view mirror), and militarily concrete was applied to creating new fortresses. Once the highly congested concrete proto-city fortresses proved of limited value, widely dispersed dirt trenches prefigured suburban sprawl, with their often erratic zigzag patterns later replicated in the 'loop and lollipop' patterns to come in post-World War II suburban developments. Concrete bunkers increasingly anchored these trench loops by opportunistically responding to and encouraging trench development, much as future suburbs would be anchored in their otherwise aimless spatial trajectories by concrete shopping malls (Funck and Chickering 2004).

Schivelbusch describes new forms of vision, such as the panoramic view, as central to the emerging stimulus shields forming around the railway. In terms of the battlefield created by rail, projectile and concrete, the most central characteristic was invisibility – the understandable subjective urge of the troops to present no outline to the murderous views of the enemy. This played out at a more systemic or objective level as well; the constant churning of the industrialized battlefield made it difficult for military commanders to identify what was happening. The solution was the aerial photograph, which could track changes otherwise impossible to register. The battlefront aerial photograph had its origins in photographs taken from balloons in the American Civil War.

Photography had long followed a developmental trajectory similar to concrete. Both initially required a cumbersome process to create the conditions necessary for them to take their final impressed shape. Both had medieval antecedents, which took recognizably modern turns in experiments done in Britain and France early in the nineteenth century. Both had contested patent claims that were bound up with technical issues, personalities, national pride and uncertainty over the emerging shape of modernity. By the middle of the nineteenth century, careful scientific analysis permitted industrial production of both technologies, but for cultural reasons both remained constrained to specific functions, such as tunnels for concrete, or the staged portrait in photography. The Great War transformed their usage in such ways that both became part of the material vernacular (Marien 1998).

Virilio argues that the battlefield use of photography was the outcome of a process that began with telescopic visual prostheses and Galileo's mathematization of nature (Johnston 1999). He cites Merleau-Ponty's description of the worldview rendered obsolete by new technologies of vision. "Everything I see is in principle within my reach, at least within reach of my sight, marked on the map of 'I can'" (Virilio 1994b, 7 quoted in Johnston 1999, 30). Although long decaying, Virilio suggests this worldview, "the age-old *act of seeing*," collapsed utterly in World War I, "replaced by a regressive perceptual state, a kind of *syncretism*, resembling a pitiful caricature of the semi-immobility of early infancy, the sensitive substratum now existing only as a fuzzy morass from which a few shapes, smells, sounds accidentally leap out ... more sharply perceived" (Virilio 1994b, 8 [orig. emphasis] quoted in Johnston 1999, 30). Virilio describes the new battlefield experience of the soldier in the trenches as 'topographical amnesia': "His faith in perception is reduced to a line of faith, the *ligne de foi*, as the gun barrel's sightline was formerly referred to in French" (Virilio 1994b, 8 quoted in Johnston 1999, 30).

A similar perceptual shift accompanied concrete's rise in the shape of the bunker. The univocal view out from the protective concrete mass became standardized in the form of urban space, especially as this developed further along its trench/suburb trajectory aided by the projectile economy. As on the battlefield, urban form reflected atomized components hurtling around with great speed, first with horse technology and more recently with trains, then multiplied by the effects of the automobile. Great concrete towers comprised of piled up bunker forms provided observation points from which the spectacle of acres of asphalt (a surfacing material embraced because of its suitability for the automobile) and all the other surrounding observation towers (LeCorbusier's 'machines for living') could be safely surveyed. The projectile economics of World War II conclusively demonstrated the merits of both concrete-hardened city centers, and flimsier structures widely dispersed in the suburbs. The concrete overpass provided the projectile economy's equivalent to the military breakthrough, each acceleration out of gridlock into the fast lane vaguely recalling the triumph of the breaking through the deadening trenches, in both cases a triumph usually more apparent and fleeting than real and permanent, despite the seemingly endless allocation of resources.

To draw toward a conclusion, it can be observed that impression has become pervasive in modernity; concrete structure, photographic images, explosive projectiles and the pounding wheels of the automobile (even more so than the locomotive) are ubiquitous. They impress themselves everywhere on the landscape, the built environment and on human bodies, whereas in the classical era examples such as the Pantheon or death masks did their impressive work more locally and tied to specific individuals. But the constitution of the self in space and time in which impressive technologies are a significant ordering principle obtains no less in the current strife in Iraq or on any local highway than in the Pantheon: forms of impression Trachtenberg summarizes (in Schivelbusch 1986) as forces whose origins remain out of view. We are pushed to ask then, beyond technical refinement, how exactly has impressive technology changed? Or, asking the same question differently, what are the consequences of impression's increasing centrality under the regime of modernity? How are these technologies of impression related to the modern self? How do they help us understand the sacrificial will-to-automobility?

The Sacrificial Self and Automobility

The important but little-known pair of lectures that Michel Foucault gave in 1980, published under the title, "Hermeneutics of the Self" (1993), help to illuminate these questions. In the first lecture, titled "Subjectivity and Truth," Foucault introduces his topic as the study of techniques of self-transformation and modification, as part of a "genealogy of the self" (Foucault 1993, 203). His first lecture focuses on Stoic notions of truth, which he characterizes as "not defined as a correspondence to reality but as a force inherent to principles and which has to be developed in a discourse" (209). In one sense this is precisely wrong: the exemplary text Foucault uses is from Seneca, which emphasizes repeatedly that truth is precisely the correspondence to reality, and that tranquility, harmony and inner peace result from the individual not contesting realities which the fates have dealt; knowledge and will are of benefit only insofar as they conform to the outline of the circumstances in which they arise. Thus Stoicism argues for a non-discursive origin of the self (Seneca 1979). But, Foucault continues, Stoic argument "is not oriented toward...individualization...by the discovery of some personal characteristics but toward the constitution of a self which could be at the same time and without any discontinuity be subject of knowledge and subject of will" (209). I think he means by this that the individual self is defined negatively by Stoicism as the sum of responses to the circumstances of fate, but the individual attains truth through the social exchange inherent in self-knowledge and will. Thus a philosophy founded on impression – an indexical correspondence between the world and knowledge, or between object and self – turns the self into a discursive formation.

Using language reminiscent of Trachtenberg's codification of Schivelbusch, Foucault argues that "[Stoic] discourse has for an objective not to add to some theoretical principle a force of coercion coming from somewhere else but to transform them into a victorious force. [Stoicism] has to give a place to truth as a force" (Foucault 1993, 209). While the true human nature revealed in Seneca's text is precisely that which conforms to exterior force – Foucault's insight that Seneca seeks truth as a force is brilliant – the force is the power of acceptance of what is, insofar as it is transmitted to the self, rather than a pointless struggle (to Seneca) over what might have been. The tranquil or true self in other words is like plaster poured into a mould of circumstance. The true self happily accepts the resultant form of itself, whatever that station in life may be. Seneca is not suggesting mere acquiescence to power. He explicitly gives as an example the warrior – a metaphor never far from the Classical mind – and warriors can after all be as easily engaged in rebellion as reactionary imperial pursuits. But whether one is a revolutionary or a dictator or swimming the vast sea of possibilities between those two poles, the self qua self that thusly confronts the world should according to Seneca's precepts accept its place and not worry about how it might have turned out differently.

With this important qualification we can accept Foucault's argument that Seneca is rejecting "a force of coercion coming from somewhere else" being added to "some theoretical principle" (Foucault 1993, 209). We can even amplify Foucault's statement: principles projected over space whose origins are impossible to see are coercive. It is precisely in this way, following Trachtenberg, that we have characterized modern forms of impressive technology in the projectile economy: the self, far from being the autonomous self comfortable in its own form that Stoics such as Seneca prescribed, has everywhere instead become the object of forces whose point of origin is out of view. So how did we get here from there? How did we get from the Stoic ideal of self – tranquil within its form, a point of view influential enough that Roman emperors could build temples to it – to the modern self overwhelmed each second by impressive forces of unknown origin? This is the topic of Foucault's second lecture, which provides a link to understanding the sacrificial will-to-automobility.

Here, clearly influenced by Nietzsche, Foucault focuses on 'Christianity and Confession'. He begins by noting that in Christianity (as opposed to say Buddhism or Gnosticism) the Augustinian 'access to the light', i.e., the light of God (or the set of propositions constituting a dogma, or the authority of the Church) is separate from the discovery of truth within the self. Whereas in other systems the two are interlinked, such that the discovery of the truth in oneself leads to enlightenment, in Christianity the two are separate, requiring two different operations. Foucault's concern in this piece is the obligation on Christians to manifest the truth about themselves (Foucault 1993).

Within Christian self-knowledge, the knowledge of the truth of the self, Foucault further bifurcates. On the one hand, there is what the Greek fathers of the Church described as exomologesis, which means showing the world the self as both essentially sinful – "dirty, defiled, sullied" – and the will of the self to "get rid of his own body, to destroy his own flesh" to get access to a new spiritual life (Foucault 1993, 214). Thus penitents wore hair shirts and covered themselves in ashes or whipped themselves raw. The ultimate expression of exomologesis is the martyr; self and self-destruction perfectly joined (215). On the other hand, Foucault continues, the Christian self requires confession, which ultimately focuses on thoughts. One might use praiseworthy actions such as fasting for sinful ends, perhaps for example to arouse the envy of other monks. The 'truth' or godliness of each moment lies therefore in how one is thinking, and thoughts must be verbalized as part of a movement toward God. Satan resists allowing his dark deeds to see the light, and thus Christians argued that verbalizations that are difficult and induce blushes or reluctance to speak indicate the devil's resistance.

True or total confession renunciates the impure self imprisoned by evil, and reaches toward the purity of God. Foucault sums up: "Verbalization is a self-sacrifice. To this permanent, exhaustive, and sacrificial verbalization of the thoughts which was obligatory for the monks in the monastic institution, to this permanent verbalization of the thoughts, the Greek fathers gave the name of exagoreusis" (220).

Foucault concludes that though these two aspects of Christian thought are very different, they are nonetheless parallel in how truth and sacrifice are closely linked. But he notes, after centuries of debate, the exagoreusis strain, the confessional technology of the self, triumphed over the exomologesis of the martyr, and is at present everywhere apparent in "the permanent verbalization and discovery of the most imperceptible movements of our self" (Foucault 1993, 222). Foucault's final thoughts are that "one of the great problems of Western culture has been to find the possibility of founding the hermeneutics of the self not, as it was the case in early Christianity, on the sacrifice of the self, but, on the contrary, on a positive, on the theoretical practical, emergence of the self" (222). This is a wonderful characterization, but overstated. Perhaps the technologies of exagoreusis and exomologesis, of the confessional and martyr, continue to run in parallel.

If we modify Foucault's conclusion thusly we are left with a rich formulation in light of our analysis of impressive technologies. It helps explain the central mystery we began with, which is the foundering of the ideal of a just, healthy and sane self (this catalogue of modalities of the self is Foucault's, and obviously reflects his research interests) on the technologies of impression. The epitome of technology of the self in the modern era has only rarely been an enlightened pursuit for a non-sacrificial self; more often the modern self has been pinned on circuits of injustice, disease, death and insanity, of which the ubiquity of automotive injury is merely one example. And at the centre of this dystopian turn in technology of the self has been speed: the thrills of the projectile economy. Schnapp (1999) characterizes speed as the unexplained motor of individuation, a psychological black box. Foucault's theory of sacrifice at the core of Christian thought helps explain it. For Christians, no greater thrill was possible than the abnegation of self through martyrdom, characterized in countless images as comparable to sexual ecstasy. This thrill underlay the quieter renunciations of the self occurring in the monastery. God, the exterior force regulating the whole system for Christians, may have died for European moderns and their successors, but the thrilling connection between truth and sacrifice remains lively. The SUV or the incendiary bomb take us to the edge of sacrifice; we who survive peer over that edge in awe.[7] The thrills of the projectile economy, and its underlying material and cultural technologies, help to explain the intractable problem of the automobile and safety, as explored in the next section. We fear and deplore the consequences, and yet there is for us

---

7   The SUV stands out as exemplary because it is specifically designed to injure and kill. See for example, Sarah Jain's discussion of causes of injury and mortality in SUV and pedestrian collisions (Jain 2004, 62; 2009, forthcoming).

no higher truth, and to it we raise concrete temples everywhere. Nietzsche would laugh; not the age of the overman, but of the overpass.

References

Aurelius, Marcus. 1992. *Meditations*. A.S.L. Farqharson, trans. New York: Knopf.

Ballard, J.G. 1973. *Crash*. New York: Farrar, Straus and Giroux.

Collins, Peter. [1959] 2004. *Concrete: The vision of a new architecture*. Montreal and Kingston: McGill-Queens University Press.

Crary, Jonathan. 1999. *Suspensions of perception: Attention, spectacle and modern culture*. Cambridge: MIT Press.

Cronenberg, David, dir. 1996. *Crash*. Alliance Communciations.

De Landa, Manuel. 1992. *War in the age of intelligent machines*. New York: Zone.

Doyle, John. 2007. Idea of blood sacrifice drowns other voices at Vimy. *Globe and Mail* 11 April: R3.

Flower, Harriet I. 1997. *Ancestor masks and aristocratic power in Roman culture*. Oxford: Clarendon Press.

Foucault, Michel. 1993. About the beginning of the hermeneutics of the self. *Political Theory* 21(2) May: 198–227.

Funck, Marcus and Chickering Roger, eds. 2004. *Endangered cities: Military power and urban societies in the era of the world wars*. Boston: Brill Academic Publishers.

Goldscheider, Ludwig. 1945. *Roman portraits*. Oxford: Phaidon Press.

Hambly, Maya. 1988. *Drawing instruments 1580–1980*. London: Sotheby's Publications.

Hirst, Paul Q. 2005. *Space and power: Politics, war and architecture*. Cambridge: Polity.

Hounshell, David A. 1984. *From the American system to mass production, 1800–1932*. Baltimore: Johns Hopkins University Press.

ICBC (Insurance Corporation of British Columbia). 2004. Traffic collision statistics: Police-attended injury and fatal collisions. http://www.icbc.com (accessed 21 August 2006).

Jain, Sarah S. Lochlann. 2004. Dangerous instrumentalities: The bystander as subject in automobility. *Cultural Anthropology* 19(1). 61–94.

———. 2009. *Commodity violence: American automobility*. Durham, NC: Duke University Press. Forthcoming.

Johnston, John. 1999. Machinic vision. *Critical Inquiry* 26(1) autumn: 27–48.

Joost-Gaugier, Christiane L. 1998. The iconography of sacred space: A suggested reading of the meaning of the Roman pantheon. *Artibus et Historiae* 19(38): 21–42.

Kern, Stephen. 1983. *The culture of time and space 1880–1918.* Cambridge: Harvard University Press.

Latour, Bruno. 1992. Where are the missing masses? Sociology of a few mundane objects. In *Shaping technology-building society: Studies in sociotechnical change,* Wiebe Bijker and John Law, eds, pp. 225–58. Cambridge: MIT Press.

MacDonald, William L. 1976. *The pantheon: Design, meaning and progeny.* Cambridge: Harvard University Press.

Marien, Mary Warner. 1998. *Photography and its critics: A cultural history, 1839–1900.* Cambridge: Cambridge University Press.

McLuhan, Marshal. 1997. *Forward through the rearview mirror.* Cambridge: MIT Press.

McNeill, William H. 1982. *The pursuit of power: Technology, armed force and society since A.D. 1000.* Chicago: University of Chicago Press.

Meeks, Carroll L.V. 1960. Pantheon paradigm. *Journal of the Society of Architectural Historians* 19(4) December: 135–44.

Rykwert, Joseph. 1976. *The idea of a town: The anthropology of urban form in Rome, Italy and the ancient world.* London: Faber and Faber.

Schivelbusch, Wolfgang. [1977] 1986. *The railway journey: The industrialization of time and space in the 19th century.* Berkeley: University of California Press.

Schnapp, Jeffrey. 1999. Crash (Speed as engine of individuation). *Modernism/Modernity,* 6(1) January: 1–49.

Seneca, Lucius Annaeus. [1935] 1979. On tranquility of mind. In *Moral essays, Volume 2,* John W. Basore, trans., pp. 202–85. Cambridge: Harvard University Press.

Swift, Emerson H. 1923. Imagines in imperial portraiture. *American Journal of Archaeology* 27(3) July–Sept: 286–301.

Transport Canada. 2005. Collisions and casualties – 1986–2005. Government of Canada. http://www.tc.gc.ca/roadsafety/tp/tp3322/2005/page1.htm (accessed 27 May 2007).

Virilio, Paul. [1976] 1994a. *Bunker archaeology.* New York: Princeton Architectural Press.

——. 1994b. *The vision machine.* Bloomington, IN: Indiana University Press.

Vitruvius Pollio, Marcus. 1914. *Ten books on architecture.* Morris Hicky Morgan, trans. Cambridge: Harvard University Press.

Wilkins, Peter. 2004. The pantheon as a globe-shaped conception. *Nexus Network Journal,* 6, 1 spring. http://www.nexusjournal.com/conf_reps_v6n1-Wilkins.html (accessed 29 November 2006).

Winter, Thomas N. 1979. Roman concrete: The ascent, summit and decline of an art. Transactions of the Nebraska Academy of Sciences. http://digitalcommons.unl.edu/classicsfacpub/1 (accessed 26 November 2006).

PART 2
Risk and Regulation

Chapter 5

The Safety Race:
Transitions to the Fourth Age of
the Automobile

David MacGregor

Introduction

"Safety doesn't sell" (Volti 2004, 116) was until recently the mantra of Detroit. Car advertisements now report IIHS or NHTSA crash test ratings, *Consumer Reports* does likewise. In exploring broad historical shifts in auto safety, this chapter shows how and why safety now sells. Building on Gartman's (2004) three cultural ages of the automobile, this chapter examines in a comparative international context changes that suggest a recent development of safety consciousness in governments, the auto industry and ordinary consumers and the influence of a safety race on automobile design. I propose that these changes constitute a fourth age of the automobile, marked by government intervention on an international scale, revolutionary advances in crash protection and avoidance technology, and consumer demand for safety features. The dialectic of motion helps to understand how these changes took place inasmuch as the safety race relies critically on the state as an intervening authority between industry and road users. Future reduction or prevention of injuries and fatalities across the world will depend heavily on government intervention in the vigorous enforcement of vehicle design that ensures the affordability of safety.

Critical to the dialectic of motion is the concept of safety consciousness. I use the term to highlight the role of human agency in the dialectic of motion, which involves the recognition of 'avoidable factors' that contribute to 'injury-producing accidents', and the willingness and ability to change these factors (Wegman 2007). As a transformational activity that takes place within a dynamic set of institutions, and social and historical conditions, safety consciousness embraces attitudes and values regarding traffic safety held by members of the general public, auto consumers, vehicle manufacturers, police officers, driving instructors, insurers, and advocacy and research organizations. Most critically, safety consciousness involves government (Hegel's 'public authority') in planning, regulating and assessing a traffic system comprised of many actors (Hauer 2007). The concept of the dialectic of motion – consisting of a complex relationship between state and civil society with respect to speed and safety – helps to explain why safety consciousness has become an integral aspect of automobility.

In drawing on Hegel (1976), Gartman argues that each auto age is motivated by a dialectical process in which contradictions growing within it are at once preserved and resolved in the succeeding age, which in turn, carries the seeds of its own eventual destruction and overthrow. While the contradictions that drive change in his theory involve class and production, Gartman also refers obliquely to a vital aspect of the dialectic of motion – the contradictions of speed and safety – that are key to explaining the safety race. He notes that mobility concerns an ultimate goal of freedom: transcendence of space and time in the rapid and secure fulfilment of multiple social connections. The three ages of the automobile encapsulate "the search for individual identity within a capitalist society that holds the promise of autonomy but simultaneously denies it in the heteronomy of the economy" (Gartman 2004, 170).

In tracing the dialectic of motion, with a focus on the contradictions of speed and safety, this chapter seeks to determine how they played out in distinct cultural ages of the automobile. I begin with an examination of safety consciousness in the first automobile age up to the 1920s and show that by the close of the second in the 1950s, a nascent safety consciousness was confronted with soaring traffic deaths and injuries. Bloody carnage on the highways in 1960s North America brought consumer activism and government intervention that characterized safety consciousness at the beginning of this third age. Both were weakened in the 1980s but their lingering influences provided a framework of state intervention for the rise of a new fourth age. Social and technological changes that emerged in the third age of the automobile produced a safety race and safety consciousness that is in the process of transforming the world of automobility. The safety race that surfaced first in the United States (less so in Canada), with the arrival of state intervention in the 1960s then shifted to Western Europe where automakers and innovative governments took the lead.

In the mid-1990s a fourth cultural age materializes in which safety consciousness becomes an integral aspect of automobility, providing a Hegelian synthesis of resolving deadly contradictions that arose in earlier ages. By late in the first decade of the twenty-first century the fourth age of the automobile has been firmly established in both North America and Europe, promising a major transformation in the system of automobility. Rather than the usual concern about changing driver behaviour or even the driving environment, government planning and intervention that focuses on improving the design of vehicles – and ensuring their affordability – are primary, if incipient and uneven, features of this age. During this period in much of Western Europe, political administrations have adopted a safety approach "that minimizes opportunities for crashes to occur and virtually precludes disabling or fatal outcomes by limiting crash severity" (Kissinger 2007, ii). Sweden's Vision Zero is one of the harbingers of the fourth age of the automobile, characterized by the safety race. A critical component of the safety race is state progress in the capacity to remove or ameliorate hazards confronted by road users. In particular, computerized vehicle systems that do not require driver intervention in vehicle crashworthiness and accident avoidance are transforming the possibilities of auto safety. On a global scale, however, the fourth auto age has not yet extended beyond high-income countries, presenting an urgent global problem.

Safety Consciousness and the Three Ages of the Automobile

The first cultural age of the automobile, lasting into the 1920s, promised a golden era of enhanced mobility but contained already the spectre of death. James Agee's (1957) Pulitzer Prize winning novel *A Death in the Family*, based on his own father's death in a single vehicle collision in 1915 when Agee was six years old, may stand as the iconic cultural statement of this fatal contradiction. Government regulation emerged simultaneously with the first auto age in North America. Driver licensing, registration of vehicles, speed laws, road building standards, and some safety provisions were in place by the 1920s (Newman 2004). However, auto manufacturing itself – by 1925 the largest US industry – took responsibility for automobile economy, efficiency and safety standards as state and federal governments mutually shrank from the task (Eastman 1984). When deaths and injury rates soared early in the twentieth century, the public grew aware of traffic dangers, but the auto industry and government ensured attention would fall most heavily on driver behaviour. Two national conferences on street and highway safety in the mid-1920s, initiated by then Commerce Secretary Herbert Hoover, cemented the notion "that the highways and automobiles were built as well as could be expected under existing technology, and that traffic accidents were therefore traceable to willful, careless, irresponsible, or incompetent drivers" (Nader 1965, 176). Compared with mass transit accidents, such as steamship explosions or airplane crashes, the perception prevailed that motor vehicle casualties resulted from shortcomings in "driver skill or care" (Mashaw and Harfst 1990, 35). What Nader called 'the safety establishment' institutionalized this perspective, using public relations methods to educate motorists about safe driving.

In the US, motor vehicle safety design made only glacial advances through most of the second age of the automobile into the 1950s. Nevertheless, fundamental changes in vehicle construction and design (not necessarily developed with safety in mind) likely had a salubrious effect, such as replacement of the crank with the electric starter, substitution of steel for wooden car bodies and the appearance of fully enclosed passenger compartments (Newman 2004). Paul Hoffman, President of the Automotive Safety Foundation (formed and funded by the Automobile Manufacturers Association), intoned in 1937, "our cars are the safest we know how to build. We will continue to build into them every sound safety factor developed by engineering genius." For Studebaker executive Hoffman, as for the rest of the automobile industry, safety was primarily about "correctible driver failures." Only "a small percentage" of accidents, suggested the safety chief, were "due to car failures" (quoted in Eastman 1984, 140–41). Moreover, "car failures," where they existed, were the result of neglectful drivers who failed to maintain and inspect their vehicles (Nader 1965, 176). During this period, Hoffman crafted the template for US auto industry (and car buyer) attitudes to safety.

A nascent safety consciousness struggled to emerge early in this second age of the automobile, with sporadic attempts to design safer automobiles in the 1930s (Eastman 1984). The road-building revolution inaugurated by Italy's autostrada, completed in 1925 and followed about a decade later by Germany's autobahn and the American freeway, had tremendous safety implications, though these were mostly submerged under the worship of speed. Just as European manufacturers lead the field today in safety technology, they set the pace following World War II (Volti 2004). British and French carmakers pioneered disc brakes in the mid-1950s. More than a decade passed before they were adopted widely in the US. Far superior in every respect to conventional bias-ply tires, radial tires developed by Pirelli and Michelin became standard equipment on most European motorcars by 1947, but did not appear on American autos until the 1970s (Volti 2004; Flink 1988).

US advances in safety technology occurred mostly on the periphery. Cars built by legendary but doomed Tucker Corporation and Kaiser-Frazer Corporation in the late 1940s and early 1950s pioneered crash protection and other innovations decades before being adopted by the major car companies. After the war, research on velocity and the human body proved that injuries in auto accidents could be reduced or eliminated by rethinking the motor vehicle and its environment. The Cornell Aeronautical Laboratory invented the crash test dummy in 1949 to evaluate the impact of simulated car crashes on the human physique, and later produced 'The Liberty Mutual Safety Car', which made headlines in 1957 (Calspan 2005). Also in 1949, Nash Motors introduced the seat belt (Newman 2004). The Ford Motor Company highlighted safety themes with its 1956 models (optional seat belts, crash proof door locks, a padded dash and deep dish steering wheel, for example), but industry worries that raising safety issues would hurt sales halted the initiative (Eastman 1984). Two-point seat belts were available in many American cars as optional equipment by 1955. Industry safety leader Volvo introduced a three-point front seat belt system in 1959. General Motors worked on airbags in the 1950s (Mashaw and Harfst 1990), decades before these devices became standard equipment in US cars (see Wetmore's chapter for a detailed history of airbags).

Safety consciousness in the United States erupted at the end of the second age of the automobile as a new consumer movement reacted to the rising tide of traffic fatalities that would pass 50,000 in the mid-1960s, heading to a peak of 56,278 in 1972 (Flink 1988). Prodded by the consumer movement, the United States government adopted an interventionist stance on auto safety starting in the mid-1960s. The safety race began in earnest with such publications as *Accident Research,* a compendium of traffic accident studies (Haddon et al. 1964) and Ralph Nader's *Unsafe at any Speed* (1965), and the hearings held by Senator Abraham Ribicoff "on vehicle manufacturer responsibility to improve auto safety" (Robertson 2006a, 183). Ribicoff's inquiry led to the Federal Motor Vehicle Safety Act of 1966. The Act established minimum crash protection standards for automobiles and created several agencies, including the National Traffic Safety

Agency (NTSA) and the National Highway Safety Agency (NHSA). Eventually the government transferred the NTSA and NHSA from the Department of Commerce to the Department of Transportation to form the National Highway Safety Bureau (NHSB), which was subsequently reorganized in 1970 into the National Highway Traffic Safety Administration (NHTSA). The 1966 Act dovetailed with a March 1968 United States Appeals Court decision holding that auto makers are "under a duty to use reasonable care in the design to avoid subjecting the user to an unreasonable risk of injury or enhancement of injury in event of a collision, which, whether with or without the fault of the user, is clearly foreseeable" (Motor Vehicle Hazards Archive Project [1968] 2006). The traffic safety agency and government more generally received critical support from a consumer movement that reached unusual prominence. Ralph Nader founded the Center for Auto Safety in 1969, staunch advocate of an effective safety consciousness.

From its birth, NHTSA faced savage attacks from the auto industry, and the unprecedented government safety initiative came to a halt at the beginning of the deregulatory, neoliberal era inaugurated by the Reagan Presidency (Claybrook 1984). As government regulation faded in the 1980s, it was replaced by renewed emphasis on driver behaviour, led by an anti-drunk driving movement that created what must be one of the most monumental efforts in social control of modern times (Reinarman 1988). Under William Haddon, US safety authorities ordered car manufacturers to install shoulder belts in 1968, and pushed for mandatory airbags, but the auto industry (which had earlier resisted seat belts) joined a rancorous debate in the 1970s and 1980s about the ineffectiveness of airbags compared to laws requiring drivers to buckle up (Robertson 2001; see Wetmore's chapter).

At the beginning of the Reagan era, political assaults torpedoed NHTSA initiatives, including the airbag and side-impact protection rules, and decimated its research capabilities. One quarter of NHTSA's workforce evaporated and a compromised NHTSA veiled its operations. Despite industry opposition, however, NHTSA "developed a complex set of rules that produced a power centre eventually involving legislative, executive and judicial branches" (Newman 2004, 222–23), forming a beachhead of safety consciousness. While NHTSA declined as a political force, the insurance industry stepped into its place. Appalled by Detroit's implacable opposition to regulation, "the insurance industry began to hold press conferences and to barrage the media with information that had been collected by the IIHS [Insurance Institute for Highway Safety]. This was to have a considerable effect in influencing the public's attitude to safety in later years and to spawn a demand by consumers for safer vehicles" (Newman 2004, 225).

In effect, forces beneath the surface were building toward eruption of the fourth age, where safety consciousness becomes integral to automobility. As the North American auto market matured by the late 1980s and globalization of the auto industry accelerated (Volti 2004), carmakers sought to differentiate their products from those of competitors. Safety technology emerged as a major focus, especially for luxury vehicles (Welch et al. 2004) in the hotly competitive North American market where Detroit's Big Three had lost their oligopolistic power to thwart safety advances. By the mid-1990s elements were in place for a dramatic transformation in safety consciousness.

Shifting into Fourth: Emerging Auto Safety Consciousness in Europe

The fourth automobile age began with remarkable transformations in safety consciousness that surfaced by the mid-1990s and rapidly accelerated in the first decade of the new millennium, including: the entrance of activist government in Western Europe; global awareness and action founded on European initiatives; significant reductions in fatalities and injuries in North America associated with previous auto crash protection and accident avoidance measures; appearance of revolutionary technological advances in vehicle crashworthiness and accident avoidance; and increased demand for auto safety, supported and stimulated by government and private advocacy organizations, among consumers in a world auto market characterized by heated competition between giant manufacturers based in North America, Europe and Asia. Thus conditions are in place for a rapid, sustained and absolute decline in automobile casualties in North America and Europe.

In Western Europe, an important manifestation of safety consciousness can be found where activist governments have taken a concerted interest in auto safety in recent years. Reflective of US safety culture during the 1960s, a new idealism characterizes safety consciousness in Europe and governments have forged a unique approach in confronting road system hazards (Forstorp 2006). In Hegel's terms, the safety idea germinated in the 1960s had returned to itself. The 1960s tumult over safety in the United States had no real counterpart in Western Europe where auto ownership rates and road traffic casualty counts were still at an early stage. Even with many fewer motor cars on the road, the European safety record was abysmal. Death rates per registered passenger vehicle in 1961 were 50 percent higher in Britain than in the US, and almost five times greater in Germany (Volti 2004, 116). By 1981, the European traffic fatality rates per 100 million miles of vehicle travel still compared unfavourably to the US (Flink 1988, 382–83). Two decades later, Sweden (8.4), the United Kingdom (7.3) and the Netherlands (8.5) surpassed the US (9.5) in fatalities per billion vehicle kilometers. Recognizing that these advances in safety were "not repeatable," the Europeans searched for new ways "to reduce the burden of road traffic casualties" (Wegman et al. 2006, 1–2).

During the early 1990s, programs in Sweden, the Netherlands and the UK caught the imagination of Europeans disenchanted with the inadequate reduction of traffic casualties. These interventionist programs highlighted the importance of car design for the prevention of casualties. Sweden's Vision Zero (Tingvall and Haworth 1999), passed by the Swedish Parliament in 1997 asserted:

> It is assumed that drivers make errors and that it is the responsibility of Swedish highway agencies to anticipate the errors and to adapt the road system to bring about the desired goal of zero fatalities and zero serious injuries. This approach

demands the long-term commitment of highway agencies, strong leadership, and a strong safety culture that can sustain the processes to achieve the long-term goal. (Bahar and Morris 2007, 373)

In seeing injury crashes as predictable and preventable, Vision Zero and the Netherlands's Sustainable Safety strategy (Wegman and Aarts 2006) view the road environment in terms of the vulnerability of the human body. Speed limits and other environmental characteristics are calculated according to the physical limits of vulnerable road users. Urban planning takes into account the safety of multiple road users with differing capabilities: e.g., children, pedestrians, cyclists and motorists.

The SUNflower (Sweden, UK, Netherlands) research initiative (SUNflower 2002) aims to produce successful national road safety strategies through data-based analysis of the comparative experience of the member countries, each of which had achieved large reductions in traffic casualties during the 1990s. SUNflower (Wegman et al. 2005) now includes nine member countries; and its influence is apparent in the groundbreaking Peden et al. (2004) study on world traffic injuries. The Swedish, UK and Netherlands safety programs also inspired an intervention-minded auto safety authority in the European Union. The EU now supports Euro NCAP (New Car Assessment Programme), an auto crash-testing program founded in the UK in 1997, that eventually gained the backing "of seven European Governments, the European Commission and motoring and consumer organisations in every EU country ... Euro NCAP has rapidly become a catalyst for encouraging significant safety improvements to new car design" (Euro NCAP 2008). Euro NCAP performs crash tests similar to those carried out by IIHS and NHTSA (but Euro NCAP does not perform rear crash tests). Unlike NHTSA, the European vehicle testing authority includes tests that measure the effect of a 40 km/h impact on pedestrians. Vehicle crash standards are developed by the EU, which issues directives on car safety design. Notably, the European Parliament passed legislation in 2003 requiring "(pedestrian) crash friendly car front[s]" on all passenger vehicles by 2015 (Wegman and Aarts 2006, 88–89). Also, the EU announced in May 2008 that electronic stability control (ESC) must be fitted on all commercial and passenger vehicles by 2014. (The EU's move followed NHTSA's announcement in April 2007 that ESC would be standard on all US autos by 2012, indicating the ongoing significance of US government intervention).[1]

In 2006, the European Traffic Safety Council carried the concept of the safety race into material existence with its Road Safety Performance Index. The Index compares traffic safety performance among EU states and recognizes best safety practices in order to stimulate political leadership. "[T]he Index covers all relevant

---

1   In contrast to a powerful regulatory presence south of the border, Canada has relied on controls over individual behaviour (mandatory seat belt laws; anti-drinking and driving legislation) rather than tackling auto design or the driving environment (MacGregor 2002).

areas of road safety including road user behaviour, infrastructure and vehicles, as well as road safety policymaking more generally" (ETSC 2008). The European Traffic Safety Council claims that improved crash safety standards could reduce deaths and serious injuries by up to 20 percent (Europa 2006). If all motor vehicles incorporated the same safety standards as those of the best cars in the same class, "half of all fatal and disabling injuries could be avoided" (Peden et al. 2004, 121). The Road Safety Index focuses mostly on driver behaviour variables, such as seat belt use, drinking and driving, and speed. However, European authorities have also established the world's first "scheme to identify roads with a high incidence of traffic accidents" (Wells 2007, 1119) – the European Road Assessment Program (EuroRap). Pioneered in Belgium, EuroRap, a sister agency to Euro NCAP, employs advanced GPS systems to prepare 'risk maps' and 'performance tracking' that are used to design safer roads.

It is difficult to say what impact new auto safety measures have had on motor deaths and injuries; trends are mixed across various countries. By 2001, EU member countries suffered approximately 50,000 motor deaths annually. The EU targeted a 50 percent reduction of traffic deaths and injuries by 2010. However, with fatalities estimated at 42,000 in 2008 (the same number as in the US), this goal will be out of reach. According to the Road Safety Performance Index (ETSC 2006), some EU countries, such as Ireland, Poland, and Hungary fared even worse than Canada and the US in reducing traffic fatalities between 2001 and 2005. However during that period, other countries such as France and Belgium substantially reduced traffic deaths (35 percent and 27 percent, respectively). Despite the difficulty in determining the causes of these diverse patterns of death rates, the fact remains that the European Union monitors these differences and seeks to establish vehicle crash standards that can serve as benchmarks for further development of the safety race.

The Fourth Age of Auto Safety in the Twenty-First Century

In tandem with government interventions in North America and Western Europe that promote vehicle crash standards, commercial interest in safety features has grown. Safety had limited commercial appeal during most of the third auto age except for a small niche market of consumers who purchased vehicles from firms like Volvo and Mercedes-Benz. But in recent years, a commercial safety race has begun to develop. By 2004, for example, mass marketer Honda sought to pre-empt Volvo in the safety field, announcing a long list of safety initiatives, including bumpers designed to avoid or reduce pedestrian injuries (Matthews 2004). Volvo responded to its various safety competitors with a promise to build an injury-proof automobile by 2020 (Edmonds 2008). Number one global auto maker Toyota's vision "is to develop 'dream cars' that are revolutionary in safety and environmental benignity" (Watanabe et al. 2007).

In the case of advertising cars, automakers may stress excitement over safety features (see Conley, this volume). But for consumers and auto industry executives, motor vehicle safety may now be as important as powerful engines and glossy car bodies (White 2006). Experts in the automotive field assert that car safety is a key concern (Perone 2007). Some Ford and Kia commercials, for example, show pickups and minivans "slamming into barriers to highlight their top ranking in government crash tests. 'You always want to hit that five-star rating'," said Kia spokesman Alex Fedorak. "Safety has to be there because cars don't sell on price alone" (Perone 2007). The fourth age flood of auto information on the internet permits consumers to choose products that emphasize safety as well as other desired characteristics (Taylor 2007).

Central to the twenty-first century safety race are recent developments in advanced technological innovations in auto design. By 1995 North American vehicle design gains in the safety race had begun to cut decisively into the road fatality rate. Equipping motor vehicles with occupant crash protection devices and other vehicle design innovations allowed the United States (and likely Canada as well) to overcome the tension between safety and an increasingly adverse automotive environment (such as higher speeds, cell phone use while driving, and vehicle crash incompatibility and increased roll-over risk associated with the rise of the SUV). Several studies provide evidence that advanced vehicle design can significantly reduce the fatal contradiction between speed and safety (see Simons' chapter on the historical dialectic of speed and collision). Farmer and Lund's (2006) study funded by the Insurance Institute for Highway Safety (IIHS) argues that declining US traffic fatalities are due to safer vehicles, not better drivers, improved roadways, or reduced speeds. The IIHS study pinpoints 1994 as the year in which driver death rates continued to decline even though speed limits increased. If changes in vehicle design had not occurred, an additional 5,000 drivers would have died in 2004. More recently, a NHTSA (2008) study indicates that auto design factors, as well as changes in behaviour, have already led to a reduction of almost thirty percent in serious traffic injuries since 1996. Similar declines in serious injuries have been recorded in the UK (Broughton and Walter 2007), where traffic deaths in 2007 fell to the lowest level since 1926 (Milmo 2008). "This tremendous progress in traffic safety improvement is likely linked," NHTSA (2008, 37) declares, to "increased use of seat belts and child safety seats; reductions in drunk driving, as well as an expansion throughout the vehicle fleet of airbags, antilock brakes, and electronic stability control" (32).

Robertson (2007) also shows that advanced technology in automobile design – including crumple zones, multiple airbags, side impact protection, static stability (anti-roll-over), and electronic stability control (discussed below) – generate large reductions in death and injury. US crash data that control for the presence of these features in fourth age vehicles produced between 1999 and 2005 showed that fatalities on US roads could potentially be reduced by 80 percent if all cars were similarly equipped. A New Zealand study that compared vehicles constructed before 1984 and cars made after 1994 (Blows et al. 2003) found that older vehicles were associated with three times more fatalities and injuries than more recent models. Robertson (2007, 310) outlines the fourth age implications of a preventive approach to traffic safety rather than the usual focus on driver behaviour by stating that, "[w]hile changing vehicles does not preclude efforts to change behaviour ... a substantial majority of vehicle-related deaths can be prevented by full adoption of changes in vehicle characteristics that are preventive, whatever the complex mix of factors that lead to serious crashes."

In particular, computerized vehicle systems are transforming the possibilities of auto safety.[2] These fourth age intelligent crash avoidance units do not require driver intervention. Unlike airbags, seat belts or crumple zones, these systems work to avoid or minimize crashes before they happen. Intelligent systems may warn the driver of an impending crash and/or directly intervene between the driver and the automobile to avoid a collision or reduce its consequences (Lie et al. 2006). The Electronic Stability Control (ESC), for example, "helps the driver to keep the vehicle under control in critical maneuvering situations" (Lie et al. 2006, 38). A Swedish study found that ESC could avert or reduce the severity of ordinary collisions by more than 25 percent. In crashes that are "more ESC sensitive, like single/oncoming/overtaking crashes on wet or icy roads, the reduction is in the order of 50 percent" (41). Except for seat belts and airbags, this fourth age technology may be the most effective safety system yet implemented. The most recent data suggests greater than anticipated benefits from ESC, with the estimated reduction of fatal crashes at 43 percent in the United States. "There were approximately 34,000 fatal crashes involving passenger vehicles in 2004. Assuming that three-fourths of these crashes involved vehicles without ESC, then about 10,000 fatal crashes could have been avoided if all vehicles had ESC" (Farmer 2006, 323).[3]

In 2005, the Mercedes E Class sedan was deemed the best performer in a safety study conducted in the United States by the Insurance Institute for Highway Safety. With ten annual deaths per million registered vehicles, the Mercedes E-Class easily surpassed the average of 87 deaths per million registered vehicles (IIHS 2005). The E-Class was one of the first vehicles to come with standard electronic stability control.[4] Two years later, the Institute reported "that all but 3 of the 15 vehicles with the lowest overall death rates have this feature, usually standard" (IIHS 2007, 7). Not only car occupants but also vulnerable road users, such as pedestrians and cyclists, can be made safer through vehicle design and intelligent systems (Peden et al. 2004). Thus the luxury class Citroën C6 features a revolutionary intelligent car design that radically improves outcomes for pedestrians in the

---

2   This chapter focuses on 'intelligent vehicles' as the key aspect of a fourth era of the automobile. However, other aspects of the driving system, especially the built environment will need to be reformed in order to conform to the outlook of Vision Zero.

3   Note that Farmer (2006) refers to fatal crashes rather than fatalities. In 2004 there were approximately 43,000 fatalities but 34,000 crashes – a ratio of 1.26 deaths per crash.

4   The Mercedes E-Class's superior safety record may be due, however, to a host of other features, for example: weight of vehicle; carefully designed crush zones; the type of people who own such a vehicle; airbag and seal belt design; etc.

event of a collision (Euro NCAP 2007).[5] In addition, smart vehicle systems can inform emergency trauma centres of the occurrence and location of an accident – especially important in "run off the road" and "decreased visibility" collisions, where stricken vehicles may not be readily seen (Champion et al. 2005, 6).

Fourth age technology, including intelligent vehicle systems that prevent crashes from happening, signal crash location and seriousness, and assist driver navigation, may be kick-started (at least in the United States) by government regulatory regimes that ensure intelligent systems are introduced in all makes of cars, luxury or not (Meckler 2006). Moreover, coordination of URGENCY software, smart vehicle systems, and emergency response and trauma centre information infrastructure will require considerable government intervention, in the form of funding, research and policymaking (Champion et al. 2005). As Beckmann suggests, the smart car – this human/machine hybrid – "eliminate[s] any sort of independence that may have been assigned to earlier stages of automobility" (2004, 89).

A significant factor in the safety race may be that automobile manufacturers focus more on high-income and better-educated consumers than in the past (Bradsher 2002). As with the Mercedes E Class, enhanced safety features often promote sales and profitability within a narrow range of luxury vehicles. Surveying the field of 'future vehicles', the IIHS reported in April 2008 that a handful of prestige automobiles, including those made by Volvo, BMW, Cadillac, Acura and Mercedes-Benz now offer intelligent systems that assist drivers in avoiding accidents and help to reduce injuries (IIHS 2008). Forward collision warning, automatic braking and lane departure warning systems are among the most promising innovations.

With safety as a saleable commodity, the auto market has, to be sure, incorporated some advanced safety features into cheaper vehicles. As we have seen, Honda took a lead in this development (White 2007). Yet, without government intervention to ensure such items are included in every vehicle, safety becomes even more a matter of affordability. "While market forces can help advance in-car safety in individual models, the aim of harmonizing legislative standards of vehicle design is to ensure a uniform and acceptable level of safety across a whole product line" (Peden et al. 2004, 120). As relatively expensive vehicles become safer, crash deaths and injuries may more often involve low- and middle-income drivers and their passengers. For instance, a 2003 US study found that motor vehicle occupant deaths vary inversely with socio-economic status measured by education – though at least some of this difference is accounted for by seat belt use, with higher SES (socio-economic status) individuals more likely to buckle up (Braver 2003). A recent report (James et. al. 2007) on preventable deaths in Canada shows that in 1996 males living in the highest quintile neighbourhoods were more likely to die

---

5   "The C6 is the first car to be awarded the maximum four stars for pedestrian protection. When the car senses that it is striking a pedestrian, a pop-up bonnet is activated, giving greater clearance between the bonnet surface and the rigid components of the engine" (Euro NCAP 2007, N.p.).

in motor vehicle accidents than the rest of the population (it was the only cause of mortality that registered higher for the highest quintile). This squares with a Norwegian study showing that wealthy people encounter a greater risk of death in auto accidents because they travel more than low and middle-income groups (ETSC 2007). Nevertheless, certain road user sub-populations that are victims of motorized transport may be concentrated among the poor, the young or the old, such as pedestrians and cyclists (Wells 2007).

The Future in the Fourth Age of the Automobile

Fourth age reductions in auto casualties may gather speed as the North American auto market switches to smaller, more fuel-efficient vehicles (Robertson 2006b). Certainly the safety gap between Europe and the US will narrow as Americans forsake light trucks in favour of crossovers (car-based SUVs) and passenger cars. For 2008, injury-producing accidents are expected to decline by 10 percent, partly because Americans drove about 3.5 percent fewer miles than the year before (autoevolution 2008) amidst the worst global downturn since the Great Depression (Wassener 2009).

The fourth age of the automobile may be only the beginning of a long transformation in which the melding of human and machine – what Beckmann calls 'the motile hybrid' – will take place within an increasingly intelligent external environment. Beckmann offers a roadmap: "On the road, hybridity translates first and foremost into two terms 'intelligence' and 'assistance'. Automotive organisms are 'intelligent' beings – they exist in 'intelligent vehicles', drive on 'intelligent streets' and unite in 'intelligent transportation systems'." (Beckmann 2004, 87). Dennis and Urry (this volume) argue that the development of 'smart cars' that operate within intelligent transport systems may produce "an epochal shift as cars are reconstituted as a networked system rather than separate 'iron cages'." Whatever the networked future of the automobile, the fourth age announces the end of a problem barely recognized at its height in Europe and North America: "automobility 'works', *because its accidents are denied*" [orig. emphasis] (Beckmann 2004, 94; Featherstone 2004). To paraphrase Marx, humankind only sets itself such problems as it can solve – the coming end of the injury-producing traffic accident discloses the real meaning of safety consciousness in the fourth age of the automobile.

References

Agee, James. 1957. *A death in the family*. New York: McDowell, Obolensky.
autoevolution. 2008. New data shows record, low highway-fatalities. http://www.
    autoevolution.com/news/new-data-shows-record-low-highway-fat alities-2644.
    html (accessed 2 January 2009).

Bahar, Geni and Morris, Nesta. 2007. Is a strong safety culture taking root in our highway agencies? In *Improving traffic safety culture in the United States: The journey forward,* pp. 367–78. Washington: AAA Foundation for Traffic Safety.

Beckmann, Jörg. 2004. Mobility and safety. *Theory, Culture and Society* 21(4/5): 81–100.

Blows, S., Ivers, R.Q., Woodward, M., Connor, J., Ameeratunga, S. and Norton, R. 2003. Vehicle year and the risk of car crash injury. *Injury Prevention* 9: 353–56.

Bradsher, Keith. 2002. *High and mighty: SUVs: The world's most dangerous vehicles and how they got that way.* New York: PublicAffairs.

Braver, Elisa R. 2003. Race, hispanic origin, and socioeconomic status in relation to motor vehicle occupant death rates and risk factors among adults. *Accident Analysis and Prevention* 35: 295–309.

Broughton, J. and Walter, L. 2007. *Trends in fatal car accidents: Analysis of CCIS data. Version 3.* Project Report PPR172. London: TRL Limited.

Calspan. 2005. Calspan corporation fact sheet. Buffalo, NY: Calspan Corporation. http://www.calspan.com/history.htm (accessed 26 January 2007).

Champion, Howard R, Augenstein, J.S., Blatt, A.J., Cushing, B., Digges, K.H., Flanigan, M.C., Hunt, R.C., Lombardo, L.V. and Siegel, J.H. 2005. New tools to reduce deaths and disabilities by improving emergency care: Urgency software, occult injury warnings, and air medical services database. Center for Transport Injury Research. http://www.cubrc.org/centir/docs/new_tools.pdf (accessed 23 July 2007).

Claybrook, Joan. 1984. *Retreat from safety: Reagan's attack on America's health.* New York: Pantheon Books.

Eastman, Joel. 1984. *Styling vs. safety: The American automobile industry and the development of automotive safety, 1900–1966.* New York: University Press of America.

Edmonds, Sarah. 2008. 2020 vision: The injury-proof car. *Globe and Mail* 1 May. http://www.theglobeandmail.com/servlet/story/RTGAM.20080501.wh-volvosafety-0501/BNStory/ (accessed 24 June 2008).

ETSC (European Transport Safety Council). 2006. Road safety performance index. Flash 6: Making progress happen. http://www.etsc.be/documents/PIN%20Flash%202.pdf (accessed 10 July 2008).

——. 2007. Social and economic consequences of road traffic injury in Europe. http://www.etsc.be/documents/Social%20and%20economic%20consequences%20of%20road%20traffic%20injury%20in%20Europe.pdf (accessed 10 July 2008).

——. 2008. The road safety performance index. http://www.etsc.be/PIN.php (accessed 12 July 2008).

Euro NCAP. 2007. European new car assessment program: Citroen C 6 rating. http://www.euroncap.com/tests/citroen_c6_2005/235.aspx (accessed 23 July 2007).

——. 2008. The official site of the European new car assessment program. http:// www.euroncap.com/about.aspx (accessed 16 June 2008).

Europa. 2006. Smart cars: Public-private drive to promote accident-avoidance technologies. Brussels: Europa Press Releases. http: //271&format=HTML &%2338;aged=0&%2338;language=EN&%2338;guiLanguage=en (accessed 26 January 2008).

Farmer, Charles M. 2006. Effects of electronic stability control: An update. *Traffic Injury Prevention* 7: 319–24.

Farmer, Charles M. and Lund, Adrian K. 2006. Trends over time in risk of driver death: What if vehicle designs had not improved? *Traffic Injury Prevention* 7: 335–42.

Featherstone, Mike. 2004. Automobilities: An introduction. *Theory, Culture and Society* 21(4/5): 1–24.

Flink, James J. 1988. *The automobile age.* Cambridge: MIT Press.

Forstorp, Per-Anders. 2006. Quantifying automobility: Speed, 'zero tolerance' and democracy. In *Against automobility,* Stefan Böhm, Campbell Jones, Chris Land and Matthew Paterson, eds, pp. 93–112. Oxford: Blackwell Publishing.

Gartman, David. 2004. The three ages of the automobile: The cultural logics of the car. *Theory, Culture and Society* 21(4/5): 169–95.

Haddon, William Jr., Suchman Edward A. and Klein, David. 1964. *Accident research, methods and approaches.* New York: Harper and Row.

Hauer, Ezra. 2007. A case for evidence-based road-safety delivery. In *Improving traffic safety culture in the United States: The journey forward,* pp. 329–44. Washington: AAA Foundation for Traffic Safety.

Hegel, Georg Wilhelm Friedrich. 1976. *Philosophy of right.* Oxford: Oxford University Press.

IIHS (Insurance Institute for Highway Safety). 2005. The risk of dying in one vehicle versus another: Driver death rates by make and model. *Status Report* 40(3) 19 March: 1–7.

——. 2007. Driver deaths: By make and model: Fatality risk in one vehicle versus another. *Status Report* 42(4) 19 April: 2–7.

——. 2008. Special Issue: Crash avoidance features. *Status Report* 43(3) 17 April: 1–7.

James, Paul D., Wilkins, Russell, Detsky, Allan S., Tugwell, Peter and Manuel, Douglas G. 2007. Avoidable mortality by neighbourhood income in Canada: 25 years after the establishment of universal health insurance. *Journal of Epidemiology and Community Health* 61: 287–96.

Kissinger, Peter. 2007. Preface. In *Improving traffic safety culture in the United States: The journey forward,* pp. i–ii. Washington: AAA Foundation for Traffic Safety.

Lie, Anders, Tingvall, Claes, Krafft, Maria and Kullgren, Aanders. 2006. The effectiveness of Electronic Stability Control (ESC) in reducing real life crashes and injuries. *Traffic Injury Prevention* 7: 38–43.

MacGregor, David. 2002. Sugar bear in the hot zone: Understanding and interpreting the political basis of traffic safety. In *Driving lessons: Exploring systems that make traffic safer,* J. Peter Rothe, ed., pp. 125–42. Edmonton: University of Alberta Press.

Mashaw, Jerry L. and Harfst, David L. 1990. *The struggle for auto safety.* Cambridge: Harvard University Press.

Matthews, Tom. 2004. Honda ad campaign takes safety seriously. *The Columbus Dispatch* 27 July. http://www.lexisnexis.com.proxy2.lib.uwo.ca:2048/us/lnacademic/re...cisb=22_T3948731957&treeMax=true&treeWidth=0&csi=1 43930&docNo=122 (accessed 26 June 2008).

Meckler, Laura. 2006. New car-safety focus: Crash prevention regulators to propose that all vehicles include stability control; weighing warning systems. *Wall Street Journal* 14 September: D1.

Milmo, Dan. 2008. Road deaths fall to record low. *The Guardian* 27 June.

Motor Vehicle Hazards Archive Project. [1968] 2006. Larsen v. General Motors Corp, United Stated of America, Motor Vehicle Hazards Archive Project. http://129.10.155.92/mvhappdfs/larsen.pdf (accessed 15 September 2006).

Nader, Ralph. 1965. *Unsafe at any speed: The designed-in dangers of the American automobile.* New York: Grossman.

Newman, Graeme R. 2004. Car safety and car security: An historical comparison. *Crime Prevention Studies* 17: 217–48.

NHTSA. 2008. *Trends in non-fatal traffic injuries: 1996–2005.* Washington: US Department of Transportation, National Highway Traffic Safety Administration.

Peden, Margie, Scurfield, Richard, Sleet, David, Mohan, Dinesh, Hyder, Adnan A., Jarawan, Eva and Mathers, Colin, eds, 2004. *World report on road traffic injury prevention.* Geneva: World Health Organization.

Perone, Joseph R. 2007. Buyers and manufacturers focusing on vehicle safety. *Newhouse News Service* 8 June. http://www.lexisnexis.com.proxy2.lib.uwo.ca:2048/us/lnacademic/re...&cisb=22_T3948548058&treeMax=true&treeWidth=0&csi=235620&docNo=14 (accessed 24 June 2008).

Reinarman C. 1988. The social construction of an alcohol problem: The case of Mothers Against Drunk Driving and social control in the 1980s. *Theory and Society* 17: 91–120.

Robertson, Leon S. 2001. Groundless attack on an uncommon man: William Haddon, Jr, MD. *Injury Prevention* 7: 260–62.

——. 2006a. Motor vehicle deaths: Failed policy analysis and neglected policy. *Journal of Public Health Policy* 27:182–89.

——. 2006b. Blood and oil: Vehicle characteristics in relation to fatality risk and fuel economy. *American Journal of Public Health* 96:1906–09.

——. 2007. Prevention of motor-vehicle deaths by changing vehicle factors. *Injury Prevention* 13: 307–10.

Taylor, Michael. 2007. Car buyers gain edge with online research. *The San Francisco Chronicle* 22 November. http://www.lexisnexis.com.proxy2.lib.

uwo.ca:2048/us/lnacademic/re...059&cisb=22_T3948548058&treeMax=true
&treeWidth=0&csi=8172&docNo=9 (accessed 2 July 2008).

Tingvall, Claes and Haworth, Narelle. 1999. Vision Zero – An ethical approach
to safety and mobility. Monash University Accident Research Centre. Paper
presented to the 6th ITE International Conference Road Safety & Traffic
Enforcement: Beyond 2000, Melbourne, 6–7 September. http://www.monash.
edu.au/muarc/reports/papers/visionzero.html (accessed 24 June 2008).

Volti, Rudi. 2004. *Cars and culture: The life story of a technology.* Baltimore:
Johns Hopkins University Press.

Wassener, Bettina. 2009. Manufacturing reports show depth of global downturn.
New York Times 3 January: B1.

Watanabe, Katsuaki, Stewart, Thomas A. and Raman, Anand P. 2007. Lessons
from Toyota's long drive, an interview with Katsuaki Watanabe. *Harvard
Business Review* July–August. http://harvardbusinessonline.hbsp.harvard.
edu/hbsp/hbr/articles/article.jsp?ml_action=get-article&articleID=R0707E
(accessed 16 July 2008).

Wegman, Fred. 2007. Road traffic in the Netherlands: Relatively safe but not safe
enough! In *Improving traffic safety culture in the United States: The journey
forward,* pp. 281–304. Washington: AAA Foundation for Traffic Safety.

Wegman, Fred and Aarts, Letty. 2006. *Advancing sustainable safety: National
road safety outlook for 2005–2020.* Leidschendam: SWOV Institute for Road
Safety Research.

Wegman, Fred, Eksler, Vojtech, Hayes, Simon, Lynam, David, Morsink, Peter and
Oppe, Siem. 2005. SUNflower+6. A comparative study of the development
of road safety in the SUNflower+6 countries: Final report. Leidschendam:
SWOV.

Wegman, Fred, Lynam, David and Nilsson, Goran. 2006. SUNflower: A comparative
study of the development of road safety in Sweden, the United Kingdom, and
the Netherlands Report. Leidschendam: SWOV. http://sunflower.swov.nl/
Reports/Paper-FW-DL-GN-AB04H413.pdf (accessed 16 July 2008).

Welch, David, Dawson, Chester, Edmondson, Gail and Rowley, Ian. 2004.
Reinventing the wheels. *Business Week* 11 October: 153. http://www.
businessweek.com/magazine/content/04_41/b3903436.htm (accessed 23 June
2008).

Wells, Peter. 2007. Death and injuries from car accidents: An intractable problem?
Journal of Cleaner Production 15(11–12): 1116–21.

White, J. 2006. Crash research: Changing fortunes of the Hyundai Sonata show
safety is about more than airbags. *The Wall Street Journal* 6 March. http://
online.wsj.com/article/SB114141689206788888.html (accessed 2 June 2006).

——. 2007. What will your daughter drive? Culling the list of used-car options
based on safety and reliability. *Wall Street Journal* 15 May: D5.

Chapter 6

Implementing Restraint:
Automobile Safety and the US Debate over
Technological and Social Fixes

Jameson M. Wetmore

Introduction

Since the 1960s automobile safety in the United States has largely been defined by an ideological dispute between two groups. One side has argued that individual drivers, passengers and pedestrians must ultimately take precautions to ensure their own safety. Those on the other side of the argument have maintained that in part because the public will never behave properly, technical fixes must be developed and mandated to decrease the risks of driving.[1] Each side has presented the government, the American public and each other with detailed strategies for improving automotive safety and backed its arguments with volumes of statistics, experiments and years of experience.

Both sides of the argument are compelling. Technical fixes have played a key role in many safety advances and often seem to be a way to simply and quickly eliminate problems. For instance the original conception of an air bag – the idea that a small box can be installed in a steering wheel (along with a handful of sensors) that will automatically deploy and protect a driver regardless of his or her disposition, conduct, or health – seems almost like a panacea.[2] But despite the allure of technical fixes, the argument that only individuals and/or social behaviour can address the immediate conditions and react to the complexities of driving (or simply walking near vehicles) is not to be taken lightly.

Of course in practice the technical and the social are never completely divorced. For instance the 'walking school buses' that Collins et al. describe (this volume) seem at first to be a clear example of a social solution to traffic safety. The main change advocated to make walking to school safer for children is the rearrangement of social structures. But physical changes must occur as well. Those setting up walking school buses might use telephones to organize their endeavours, they may set up signs in an attempt to calm traffic, and may increase enforcement of speed

---

1   See for instance Haddon (1974).

2   MacGregor (this volume) makes a similar argument for electronic stability control systems.

laws. The results of their efforts might decrease the number of cars on the road and perhaps the routes that people drive. John Urry (2004) explains this type of interconnectedness well when he describes the 'system' of automobility.

But even if approaches to automobile safety must almost necessarily combine social and technical elements, they are often defined as one or the other because these labels connote an important difference in how responsibilities are distributed. The arguments about automobile safety are not simply about differing theoretical analysis, they are also about responsibility – about who will be responsible for the money, time, and effort required to make the roads safer, and who will be held legally liable for injuries on the road (Wetmore 2004). Different groups have aligned themselves not only with different ideologies, but with conceptions of responsibility that correspond with those ideologies. Typically automakers and suppliers have argued for putting the onus on individual drivers and resisted the idea that government regulations should require corporations to accept new responsibilities.[3] Organizations traditionally focused on protecting individuals – including the National Safety Council, insurance companies, and groups founded by Ralph Nader like Public Citizen and the Center for Auto Safety – have strongly argued that car companies must be closely regulated to ensure that they take auto safety seriously and build 'responsible automobiles'.

The goal of this chapter is neither to explain these politically charged arguments in detail nor to come down on one side or the other of them. Rather it is to explore the problems that are generated by such disputes and, more importantly, to demonstrate the benefits to all parties, and safety in general, of breaking from close-minded ideologies and developing and accepting not only multiple approaches, but also a distribution of responsibilities (Wetmore 2007). At various times over the past forty years, through a variety of political techniques, one of the two ideological camps has gained an upper hand and was able to implement its strategy to some degree. Some of these efforts met with success; others failed miserably. For the most part these disputes led to standoffs during which time very little was accomplished. On a few occasions, however, those most interested in implementing technical fixes and those convinced that social change was the only solution set aside their differences, began to work together to implement both strategies simultaneously, and thereby reinforced one another's efforts. This chapter uses one episode in the history of automotive restraints to analyze the rationale behind the different

---

3   The Collins et al. chapter provides an example of how efforts to increase automobile safety are often framed by specific ideas about responsibility. The walking school buses they describe are at least in part motivated by the fact that in New Zealand pedestrians, rather than drivers, are typically blamed when pedestrians are struck by vehicles. Peter Norton (2007) demonstrates how this sort of distribution of responsibility is a political achievement. He chronicles how a number of organizations and corporations went to great lengths to convince the American public (and judges) in the first few decades of the twentieth century that pedestrians, not motorists, were at fault when they were injured or killed while crossing the road.

strategies for distributing responsibilities for automobile safety and to examine how each fared on its own and in concert with the other.

Seat Belts as Automotive Restraints

During the 1940s and 1950s a number of researchers and doctors, like Col. John P. Stapp and Hugh DeHaven, argued that the dangers of automobile travel could be significantly alleviated if occupants were properly restrained inside the vehicle.[4] The researchers claimed that there was room inside vehicles for people to survive even high speed crashes. To make this idea work, however, the occupants somehow had to be prevented from violently impacting the interior of the automobile and, even more importantly, prevented from flying out of the automobile during a collision. The solution they developed was the seat belt. The safety belt, they argued, could distribute the forces of a crash more evenly across a person's body and make crashes much more survivable.

By the mid-1950s most major American automobile manufacturers responded to this idea by offering seat belts as an extra cost, dealer-installed accessory on some of their cars.[5] In the early 1960s, governments in the United States became increasingly involved in promoting the seat belt; a handful of states began mandating that passenger vehicles sold in their state be equipped with lap belts for front seat occupants and the federal government issued a similar requirement for the fleet vehicles it purchased (Bingham 1973, 32). As more and more of these regulations were passed, the US automobile companies found that it was more cost effective and easier to manage their product by simply equipping all the vehicles they produced with belts. So on 1 January 1964 they made front lap belts standard equipment and in the 1966 models they included standard rear lap belts as well (Graham 1988, 32; Hill 1967, 202). To ensure that automakers continued to install the technology, the National Highway Safety Board (the Federal Agency charged with regulating motor vehicles to make them safer) issued Federal Motor Vehicle Safety Standard 208 which required that all vehicles sold in the United States be equipped with lap belts and shoulder harnesses for front seat positions and lap belts for back seat positions beginning 1 January 1968 (US Department of Transportation 1967).

Despite the widespread availability of seat belts by the late 1960s, the technology alone did not ensure that automobile occupants would be properly restrained. A seat belt that was not properly fastened provided no benefits whatsoever. Thus

---

4   For instance, the American College of Surgeons first recommended in February 1954 that automobile manufacturers include seat belts in all vehicles (Eastman 1984, 195).

5   Nash began offering lap belts in 1950. Ford and Chrysler began in 1955. GM followed suit in 1956 (Eastman 1984, 227; Motor Vehicle Manufacturers Association of the United States 1974, 123).

the second half of the seat belt strategy was to develop social mechanisms that would convince drivers to use the devices. A number of organizations stepped in to do this. The American Medical Association, the US Public Health Service, and many of the organizations that had traditionally been involved in driver education (like the American Automobile Association) published flyers, aired public service announcements, and spoke to local groups, in a campaign to convince Americans to buckle up (US Senate 1965–1966, 1124–1162). Automakers explained in owner's manuals how the belts were to be used and why.

Developing an Alternative

Despite all these efforts, by the early 1970s an increasingly vocal group of organizations including the National Highway Safety Board (which became the National Highway Traffic Safety Administration in 1972), insurance companies like Allstate, and safety organizations like the National Safety Council were seriously questioning the effectiveness of this seat belt strategy. They argued that the approach was fundamentally flawed because it had unrealistic expectations of the American public and overestimated the possible effectiveness of using social strategies to convince individual drivers and passengers to change their behaviour. As the National Highway Safety Board put it in 1970: "Experience since the issuance of the initial safety standards has shown that public resistance to the use of seat belts prevents them from achieving their potential for reducing the death and injury rate" (NHSB 1970a, 7187).

The best way to ensure automotive restraint, they argued, was to place less emphasis on attempting to change the public's habits and more emphasis on developing a technological fix. They contended that it was not possible to make the general public behave responsibly and convince them to wear their seat belts every time they got into their automobiles. To successfully address the problem of restraint, they suggested that safety advocates focus on automotive design.

The solution they most heavily promoted was the automobile air bag. The idea was that air bags installed in every new car sold would automatically deploy in the event of a collision and provide a cushion of protection for all occupants. Instead of relying upon millions of individual motorists to buckle their belts on a daily basis, air bag advocates promoted a solution that they believed would not require such discipline.[6] They predicted that once the devices were installed, the problem of restraint would be solved.

The air bag advocates were enamoured with the idea because it was a technical fix; it was a solution that they thought could be developed by a handful of engineers, executives and regulators, and then distributed in an easily packaged form to all automobile occupants. As one life-long air bag promoter, William Haddon, Jr, former head of the NHSB and then president of the Insurance Institute

---

6   Bruno Latour offers a number of instances where such an approach was taken in his famous "Missing Masses" article (1992).

for Highway Safety, stated in 1974: "The universal provision of air bags to achieve [automotive restraint] … would require only … a simple binding decision, by one federal official or by some three or four executives of motor vehicle manufacturing companies" (Haddon 1974, 9). Air bag advocates in the 1970s argued that the technological fix could bypass the contingencies of relying on individuals and ensure proper restraint for all.

But this strategy was not as simple as building new devices and installing them in automobiles. It would require a rather marked redistribution of responsibility and an accompanying complex set of social, political and legal changes. Most crucial to the success of the plan were the changes being demanded of automakers. Previous to the air bag argument, they were only responsible for providing a reasonably simple technology – the seat belt. The ultimate onus fell on the shoulders of motorists who had to faithfully use the device. The air bag plan required much more of the automobile manufacturers. Under the plan they did not share responsibility for restraint with motorists – they assumed practically all the responsibility. Federal Regulators went so far as to say that a vehicle equipped with air bags need not even be equipped with seat belts (NHSB 1970b). This idea of restraint put a brand new and vastly expanded responsibility squarely on the shoulders of automobile manufacturers. This new responsibility would require them to devote millions of dollars as well as considerable labour and time to develop, refine, build and install a new and complicated technology. But perhaps more importantly it also opened them up to a vast array of legal liabilities. If automakers were responsible for providing protection in each and every crash, in some sense they were at least partly responsible for each and every injury and death that occurred in automobiles. It was conceivable that automakers could be sued for not only the 50,000 plus fatalities that occurred every year in the United States at the time, but the countless injuries as well.

The U.S. automakers had little desire to accept costly changes that could possibly open the door to a huge new set of liabilities and for the most part they strongly resisted the push for air bags in the late 1960s and 1970s.[7] Instead they insisted that automotive design could never completely solve the problem of restraint. They maintained that when properly used, the seat belt was the most effective restraint device available. Because air bags were designed only for frontal crashes and provided no protection in rear impact, side impact, and rollover crashes, automakers were increasingly frustrated with what they believed were inflated promises of such a technological fix. If the belief that air bags were a panacea was widespread, a huge number of automotive dangers would be ignored (and of

---

7   The one US exception to this rule was General Motors for a brief period during the early 1970s. During that time GM's president Ed Cole strongly backed air bags and had his engineers develop and install the technology in about 10,000 Chevrolets, Cadillacs, Buicks, and Oldsmobiles (Jones 1971; Johnston 1997; 2000, 189). When Cole retired in 1974, however, the support that air bags had from the top disappeared and GM air bag production ceased altogether by 1976.

course lawsuits would be filed against corporations for injuries that their technology could not possibly have prevented). Seat belts, on the other hand, when fastened as they were designed, provided substantial protection in all types of collisions. Thus automakers (and other organizations including AAA) argued that for the good of all motorists, air bag advocates needed to stop distracting the automotive safety community from the most important approach to safety and the need to convince automobile occupants to use the existing technology, the seat belt.

A Standoff over Restraints

This resistance led to a standoff. Automotive safety regulators in the federal government had begun a concerted push to mandate air bags in all new automobiles beginning in 1968 (Graham 1989). But through the 1970s and early 1980s, these efforts never resulted in a regulation that stayed on the books long enough to be implemented. Continued resistance from most automakers and changes in government administrators prevented the air bag mandate from being implemented.

For instance, on 3 November 1970, Douglas W. Toms, Director of the National Highway Safety Bureau, issued a new government regulation which essentially mandated air bags in all passenger cars by 1 July 1973 (NHSB 1970b).[8] In response, Chrysler, Jeep, American Motors, Ford and the Automobile Importers of America took the federal regulatory agency (which by then had become the National Highway Traffic Safety Administration or NHTSA) to court arguing that the technology the agency was demanding was not feasible. The 6th US Circuit Court did not give the automakers an absolute victory, but did decide that the regulation did not adequately establish specifications for the crash test dummies required in the tests and required NHTSA to postpone its regulation until it developed such specifications (Chrysler v. DOT 1972).

In 1974, NHTSA announced that the dummy specifications were reworked such that they satisfied the court's demand and issued another proposal for passive restraints to begin in 1977 (NHTSA 1974). But the automakers continued to resist. Every time the NHTSA Administrator or the Secretary of Transportation was replaced (which happened at least every time a new president was elected), the restraint regulations were revisited and the automakers had another chance to lobby for changes. Thus when Gerald Ford appointed a new Secretary of Transportation in 1975, the 1977 deadline was re-evaluated and pushed back (US Department of Transportation 1976). Under Jimmy Carter, NHTSA issued a mandate that would begin in 1981 (NHTSA 1977), but soon after Ronald Reagan was elected, his

---

8   Technically the rule stated that the vehicle must be able to have safety equipment that will protect 50th percentile adult American male dummies in a 30 mph crash into a wall without requiring the occupants to take any action. This was to be accomplished through devises called 'passive restraints' because they provided restraint while allowing the automobile occupants to remain passive. At the time, the most widely discussed passive restraint was the air bag.

administration revoked the Carter regulation (NHTSA 1981). Between automaker resistance and the changing politics in Washington, DC, the air bag advocates could not rally enough support to implement their strategy.

As the debates over air bags raged, those who argued that the public must be convinced to wear seat belts made a handful of modest efforts to do so. They attempted to spread the message through television commercials, print ads and instructions on how to use them in owner's manuals, etc. But these small projects had little to no effect. Throughout the 1970s seat belt use in the US remained somewhere between 3 percent and 10 percent (NHTSA 1992).

With both sides of the argument mired in failed attempts to implement their strategies independently, very little progress was made during the 1970s. Much of the efforts by those interested in automobile safety were focused on the dispute over how to best protect motorists, rather than on taking concrete steps that actually would protect motorists. The automakers and their allies insisted that automobile safety could not be significantly furthered if motorists did not take an active role in the process. Government officials, insurance companies, and others insisted that technical solutions could be found to all of these problems if only the automakers accepted their 'obvious' responsibilities. These divisions and distractions created new problems and meant that very little progress was made on directly addressing the problem of deaths and injuries on the nation's roads.

Redefining Air Bags and the Distribution of Responsibility

After almost two decades of disputes, in the mid-1980s this standoff between advocates of a technical fix and advocates of a social fix to automotive restraint began to weaken. Two major events that happened in 1984 helped make possible a more cooperative 'third way' to address automobile safety.

The first event was a push by Mercedes-Benz for a new automotive restraint strategy. The German car company began offering air bags in Europe in the early 1980s and sought to bring the technology to the US in 1984. Its executives did not, however, want to be saddled with the expectation that its air bags would or could provide complete restraint for motorists. They were afraid of NHTSA portraying air bags to the public as a safety cure-all or as a restraint that would free the public from all responsibility and place the onus on automakers. Therefore, Mercedes engineers and administrators sought to reinvent the air bag – not through technical changes, but rather by redefining its use, by explaining the social change that should accompany the new technology.

When they introduced their air bag in the United States they argued that it did not replace the seat belt – it supplemented the seat belt. They called their air bag a 'supplemental restraint system' and stressed the importance of motorists wearing their safety belts – calling the devices 'indispensable'. The importance of this message can be seen in the fact that Mercedes executives did not simply slip it into an owner's manual that few people read. They actually described the use of air bags and the importance of wearing seat belts in an advertisement that ran in widely read magazines like *Newsweek* (Mercedes-Benz 1984). This two page, text-intensive spread is a very rare instance where an automobile advertisement doesn't brag about the product as much as it explains the limitations of the product and responsibilities of the owner.

In essence, Mercedes-Benz's goal in publishing this advertisement was to redefine how responsibility for restraint could be distributed. The company argued that restraint should be achieved both through technical actions and the actions of motorists. While the Mercedes idea wasn't entirely new – it certainly had been discussed in professional and technical meetings on automobile safety – this advertisement was the first time it was clearly presented to the American public as a way to frame automobile safety. The Mercedes approach was not to take a side in the air bag/seat belt standoff, but rather to argue that both technologies should be used simultaneously and that efforts to provide effective restraint should rely on techniques that heavily involved both social and technical means.

The presentation by Mercedes-Benz did not fall on deaf ears. Other manufacturers immediately recognized the value of the approach and adopted it as their own. Reframing the air bag as a supplement to – not a replacement of – the seat belt made installing the technology a significantly more palatable idea to automobile companies. Presenting air bags as supplementary restraints limited the responsibilities that manufacturers would have to accept, maintained the idea that automobile occupants still had a large role to play in their own safety, and thus eliminated many of the liability concerns. By 1993 at least eight other major automobile manufacturers used the same exact phrase – Supplemental Restraint System – to describe the airbags in their vehicles.[9]

Mandating Seat Belt Use

The second important change in 1984 was initiated by the US federal government. After years of frustration with not being able to keep an air bag regulation on the books due to the resistance of automakers and their allies, Elizabeth Dole, the second Secretary of Transportation under Ronald Reagan, sought to break the air bag impasse by developing a unique solution.

In July 1984 Dole issued a revised 'automatic restraint' regulation that required that all cars offer protection by 'means that require no action by vehicle occupants' for frontal and front angle crashes, to be phased in by 1 September 1989 (NHTSA 1984). In practice this regulation would require that automobile manufacturers install a technical fix in all their automobiles – either air bags or automatic safety belts (belts that automatically strap occupants in). This regulation was like the many failed regulations that came before it except that

---

9   Honda, Nissan, Toyota, Mitsubishi, Ford, Saab, BMW, and Jaguar have used the phrase in their advertisements and owner's manuals.

Elizabeth Dole added what became known as a 'trapdoor'. The regulation stated that if the Secretary determined by 1 April 1989 that two-thirds of the nation's population were covered by state mandatory seat belt use laws that met certain requirements, the automatic restraint regulation would be dropped. Instead of choosing one side of the standoff over the other, Dole acknowledged that both sides had their benefits. The design of her regulation was based not on an ideology, but rather on the goal of making greater strides toward increased automotive restraint – whether through social means (mandatory safety belt use laws) or technical means (automatic restraints).

The regulation had the effect that Elizabeth Dole desired – it shifted the efforts of those involved in the standoff from lobbing accusations at each other to taking concrete steps to increase automobile safety. The most marked change was made by the automobile manufacturers. They jumped on the possibility of avoiding air bags. Instead of confining their seat belt promotion efforts to a page of their owner's manuals and the occasional brochure, they formed a coalition they called 'Traffic Safety Now' and launched the largest seat belt advocacy campaign ever developed. Through the second half of the 1980s, automobile manufacturers would spend over 100 million dollars in an effort to increase seat belt use (Traffic Safety Now 1992, 118).

Traffic Safety Now took two simultaneous approaches. First, it lobbied state governments for mandatory seat belt use laws. These laws would make it illegal for front seat occupants to travel without a belt properly fastened.[10] But the entire auto safety community knew that it would take more than laws to change public practices and convince Dole to repeal the air bag regulation. Therefore, second, it put a great deal of effort into educating the public on the importance of wearing seat belts. Traffic Safety Now, along with the National Safety Council, NHTSA, insurance companies, and others promoted seat belt use through brochures, outreach to elementary schools, programs with police officers, and public service announcements like the Vince and Larry crash test dummy television commercials. This effort was quite different from the small, haphazard education programs of the 1970s. It was undertaken by a number of groups in a well coordinated effort that saturated the American public – from grade schoolers to the elderly.

---

10   Initially all the laws passed in the states were what are now known as 'primary enforcement laws'. That means that individual motorists could be pulled over and fined simply for failing to wear a seat belt. This generated some controversy in African American communities who did not like the idea of giving police officers one more reason to pull over black motorists. Thus a number of states passed what are now known as 'secondary enforcement laws'. Under such laws motorists cannot be pulled over simply for not wearing a seat belt, but if they are pulled over for another reason, they can receive an additional fine for failing to buckle up.

Simultaneously Pursuing Social and Technical

The concerted effort in the 1980s to pass mandatory use laws and increase seat belt use resulted in massive changes. First of all, mandatory use laws were passed across the country. In 1983 only New York State had such a law. By 1990 such laws were in effect in thirty-eight states (NHTSA 1996, 39). Today front seat occupants are required to buckle up in all states except New Hampshire. Second, the combination of these laws and the widespread efforts to educate the American public had a huge impact on the number of people wearing their safety belts. In 1984 NHTSA studies estimated that about fourteen percent of adults wore their seat belts in the front seat. By 1991 it had jumped to nearly sixty percent (NHTSA 1992, vii).

The third major result of these efforts was more of a side effect. The fact that more and more people were wearing their seat belts helped to pave the way for a more widespread acceptance of air bags. As was already mentioned, a number of automobile manufacturers latched onto Mercedes Benz's idea of describing (and designing) air bags as supplemental restraints. But initially they struggled to spread this idea. NHTSA and other air bag enthusiasts had once stressed air bags as the primary restraint for an American public that refused to wear seat belts and it was difficult even for car manufacturers to reframe the meaning of the technology. But as belt use increased, NHTSA officials and organizations like the National Safety Council and the Insurance Institute for Highway Safety became more comfortable with the idea of air bags as a 'supplemental restraint'. Since the mid-1980s – and increasingly so to this day – NHTSA has stressed the importance of the air bag-seat belt combination to the public through outreach programs like its 1988 public education film, *The Winning Combination* (NHTSA 1988). By the early 1990s NHTSA referred to air bags as supplemental restraints just as the automakers did.

As air bag advocates became more comfortable with the idea that seat belts could be useful, automakers became more comfortable with air bags. The automaker interest in air bags was piqued both by the fact that NHTSA began to relax the idea that air bags must provide all restraint in a collision and by the fact that as more Americans buckled up, less would be required of the air bags. These changes significantly decreased the automakers' fears of liability.

In spite of the success of the efforts to increase seat belt use, the Dole 'trapdoor' did not go into effect.[11] Starting in 1989, automobile manufacturers

---

11   There are a number of reasons why the trapdoor did not go into effect but the most important factor was a decision made in California. Those who wanted to ensure that air bags became mandatory sought to close the trapdoor by preventing such a law from being passed in California because it was the nation's most populous state and it would be almost impossible to get two-thirds of the American population covered by mandatory use laws without it. They found an ally in Willie Brown, then Speaker of the House of Representatives of California. Brown agreed that air bags were important, but he did not

were required by NHTSA to begin phasing air bags into their automobiles. But by the time the regulation went into effect automakers did not mind so much. In the late 1980s, automakers did not just abide by NHTSA air bag regulations – they were installing the technology ahead of the mandate and making it part of their marketing campaigns.

The increased interest in automobile safety generated at least in part by the debates over and education about the mandatory seat belt use laws had convinced the automakers that people would purchase a particular vehicle just because it had air bags. The manufacturers came to believe that safety would sell (MacGregor, this volume). Thus Lee Iacocca loudly proclaimed in 1988 that air bags were standard equipment on six of Chrysler's vehicle lines and that all domestically produced Chrysler vehicles would have standard air bags by 1990 (Chrysler Motors 1988; Plymouth 1991). In 1987 General Motors executives issued a press release describing why their decade-long opposition to air bags was coming to an end:

> The market now seems ready to accept air bags as a supplement to manual lap/ shoulder belts. We also were motivated by what we perceive to be a more positive attitude today concerning air bag technology, where the inflatable restraint is used in conjunction with belts. We believe that supplemental air bag technology has come of age. (General Motors 1987)

After decades of fighting the technology, automakers began producing airbags as quickly as they could.

By the early 1990s a majority of Americans were wearing their seat belts and automakers were installing airbags at an ever increasing rate. The two camps into which the auto safety community had divided were not necessarily working hand-in-hand, but they were contributing to a common and integrated project. The idea that all efforts should be placed on either a social fix or a technological fix had been abandoned. The groups simultaneously carried out the two approaches.

Conclusion

Highway safety has been a pressing issue in the United States for about a century. Every year over 40,000 people are killed on the nation's roads (NHTSA 2005). Countless groups have proposed hundreds of different strategies to remedy this

---

want to abandon the momentum that was pushing for a mandatory belt law either. To solve this problem he put a 'poison pill' clause in the mandatory belt use legislation which stated that if the federal government ever tried to use the California population to trigger the trapdoor, the California law would immediately disappear from the books. This move effectively ended any chance that mandatory use laws could pre-empt the air bag regulation (Hurley 2000; Wolinksy 1985).

problem. Some of these strategies have been largely characterized as a social fix while others rely largely on a technology to resolve the problem.

There can be a great temptation to think that if a social approach fails, it must be replaced with an approach that relies much more on technology. Likewise, when technologies seem to have little effect, it seems that they should be scrapped altogether. But when these approaches are presented as diametrically opposed to one another, two things happen. First, the benefits of combining the approaches tend to be ignored. And second, groups start taking sides in the debate over strategies and lose track of the problem they are supposed to be solving.

Early efforts to increase automotive restraint in the United States were largely hamstrung by the debate over how to best achieve it. Those advocating the social fix of convincing the American public to buckle their seat belts stood in loud opposition to those who pushed for airbags as a technological fix, and vice versa. As a result a large amount of the people and resources that could have been directly addressing safety were focused on the political debates. Only once both sides of the argument began to redefine the opposing strategy as a complementary strategy did either side have any significant effect on the driving public. In the 1980s both sides gradually came to this conclusion and for the first time a widespread effort to push both social and technical fixes was realized.

The history of auto safety has shown that when key parties build walls and file court cases against each other, an atmosphere of distrust develops and efforts to improve automobile safety are stalled. In automotive restraint some of these walls have been torn down, but other distracting disputes remain. NHTSA and the automakers are currently debating what sort of electronic safety equipment should be mandated, including devices that help to stabilize vehicles to prevent roll-overs and newly developed airbags that sense the size of specific passengers and tailor deployment to be both safe and effective for them. The safety race between automakers, as MacGregor's chapter suggests, may accelerate over the next decade. But disputes over strategies and responsibilities can create a dangerous situation if they become heated and lead to different groups creating hard stances against other approaches. When energies are placed on battling an opponent, the problem that both sides are trying to solve often gets ignored.

The problem of automobile safety is too great to focus on single solutions or to blame a single institution or group as the cause of the problem. To enact real change, those engaged in enhancing the safety of automobile travel must recognize the interconnectedness of technologies, individuals and institutions (Urry 2004). The airbag debates have shown that forcing all responsibility for the problem onto the public, the automotive industry or even the government, is not only ineffective but can distract from other solutions and strategies. Multiple strategies need to be employed and responsibility for addressing automobile safety must be distributed widely to ensure a greater possibility of success.

References

Bingham, M.P. 1973. *Safe driving is no accident*. New York: Cambridge Book Company.

Chrysler Motors. 1988. No one can guarantee safety. But that doesn't mean the car industry shquldn't keep trying to do more. Advertisement. *Newsweek* May 30: 18–19.

Chrysler v. DOT, 472 F.d2 659 (6th Cir. 1972).

Eastman, Joel. 1984. *Styling vs. safety: The American automobile industry and the development of automotive safety*. Lanham, Maryland: University Press of America.

General Motors. 1987. *Press Release* 5 March. Detroit, Michigan.

Graham, John D., ed. 1988. *Preventing automobile injury: New findings from evaluation research*. Dover, MA: Auburn House Publishing Company.

———. 1989. *Auto safety: Assessing America's performance*. Dover, MA: Auburn House.

Haddon, William, Jr. 1974. Passive vs. active approaches to reducing human wastage. *IIHS Status Report* 9(19) September: 9.

Hill, Frank Ernest. 1967. *The automobile: How it came, grew and has changed our lives*. New York: Dodd, Mead and Company.

Hurley, Chuck. 2000. Director of the air bag and seat belt safety campaign and former head of government relations, National Safety Council. Interview, 28 June.

Johnston, James D. 1997. *Driving America: Your car, your government, your choice*. Washington: The AEI Press.

———. 2000. Former vice president for Government Affairs, General Motors Corporation and former Traffic Safety Now board member. Interview, 9 November.

Jones, Stacy V. 1971. G.M. President receives air bag patent. *New York Times* 9 October: N.p.

Latour, Bruno. 1992. Where are the missing masses? The sociology of a few mundane artifacts. In *Shaping technology/building society: Studies in sociotechnical change,* Wiebe E. Bijker and John Law, eds, pp. 22–58. Cambridge: MIT Press.

Mercedes-Benz. 1984. The Mercedes-Benz supplemental restraint system: It works slightly faster than you can blink an eye. Advertisement. *Newsweek* 23 April: 45.

Motor Vehicle Manufacturers Association of the United States. 1974. *Automobiles of America: Milestones, pioneers, roll call, highlights*. Detroit: Wayne State University Press.

NHSB. 1970a. Occupant crash protection; Passenger cars, multipurpose passenger vehicles, trucks and buses: Notice of proposed motor vehicle safety standard. *Federal Register* 35(89) 7 May: 7187–89.

———. 1970b. Motor vehicle safety standards: Occupant crash protection in passenger cars, multipurpose passenger vehicles, trucks and buses (Final Rule). *Federal Register* 35(214) 3 November: 16927–31.

NHTSA. 1974. Motor vehicle safety standards: Occupant crash protection (Notice of Proposed Rulemaking). *Federal Register* 39(54) 19 March: 10271–73.

———. 1977. Federal motor vehicle safety standards: Occupant restraint systems (Final Rule). *Federal Register* 42(128) 5 July: 34289–99.

———. 1981. Federal motor vehicle safety standards: Occupant crash protection (Final Rule). *Federal Register* 46(209) 29 October: 53419–29.

———. 1984. Federal motor vehicle safety standard; Occupant crash protection (Final Rule). *Federal Register* 49(108) 17 July: 28962–9010.

———. 1988. *The winning combination*. Public Education Film. Washington, DC.

———. 1992. Evaluation of the effectiveness of occupant protection: Federal motor vehicle safety standard 208. Interim Report, June.

———. 1996. Effectiveness of occupant protection systems and their use: Third report to Congress. Washington, DC, December.

———. 2005. Traffic safety facts: 2004 traffic safety annual assessment: Early results. Washington, DC, August.

Norton, Peter. 2007. Street rivals: Jaywalking and the invention of the motor age street. *Technology and Culture* 42(2): 331–59.

Plymouth. 1991. Think of it as a mobile life support system. Advertisement. *Newsweek* 18 November: 9.

Traffic Safety Now. 1992. *An American revolution: The story of Traffic Safety Now*. Washington, DC.

Urry, John. 2004. The 'system' of automobility. *Theory, Culture & Society* 21(4/5): 25–39.

US Department of Transportation. 1967. Initial federal motor vehicle safety standards; motor vehicle safety standard no. 208: Seat belt installations – passenger cars. *Federal Register* 32(40) 1 March: 3390.

———. 1976. The Secretary's decision concerning motor vehicle occupant crash protection. Washington, DC, 6 December.

US Senate. 1965–1966. Traffic safety: Examination and review of efficiency, economy and coordination of public and private agencies' activities and the role of the federal government: Hearings before the Subcommittee on Executive Reorganization of the Committee on Government Operations. Eighty-Ninth Congress, First and Second Sessions, 22, 25–26 March 1965; 12–13, 15, 21 July 1965; 2–3, 10 February 1966; 22 March 1966.

Wetmore, Jameson M. 2004. Redefining risks and redistributing responsibilities: Building networks to increase automobile safety. *Science, Technology and Human Values* 28(3): 377–405.

———. 2007. Distributing risks and responsibilities: Flood hazard mitigation in New Orleans. *Social Studies of Science* 37(1): 119–26.

Wolinsky, Leo C. 1985. Brown holds out a compromise to pass seat belt bill. *Los Angeles Times* 5 September: 3.

Chapter 7

'Mind That Child': Childhood, Traffic and Walking in Automobilized Space

Damian Collins, Catherine Bean and Robin Kearns

Introduction

Children loom large as a group who may be marginalized and placed at risk by automobility. While the costs of automobile dependence and traffic are by no means evenly distributed among children, certain broad patterns can be observed. One such pattern, documented in early critical work on the social construction of traffic safety, is a tendency to blame the victim when children are injured by motor vehicles. Such thinking, while flawed, underscores programmes that seek to modify the behaviour of children – rather than the priorities of automobile-dominated urban environments – in order to reduce the risk of injury. This study seeks to extend existing scholarship on the often precarious place of children in auto-dominated urban environments. It does so with particular reference to the views of parents who walk with groups of children to and from primary (elementary) schools in Auckland, New Zealand. Drawing on three years of walking school bus (WSB) survey data, it notes that, in a heavily trafficked environment, parents sometimes construct children as risky and unpredictable 'objects'. Parents' discourses are more complex than this, however: they also critique the auto-dominated priorities of the city, and acknowledge children as a legitimate presence on the street, who enjoy and benefit from active travel. Parents' perceptions of child pedestrian safety also invoke particular understandings of gender identity (such as boys as boisterous, unruly pedestrians).

This chapter proceeds as follows. First, we provide an overview of the effects of automobility on the social construction of childhood, and on the organization of family life more generally. We contend that auto-dominated urban space, in particular, has played a central role in constituting and mediating particular social relations across generations and between genders. We then introduce the Auckland context and our ongoing research into children's active travel in this automobile-dependent city. The following results section focuses on the views of adult survey respondents concerning the benefits and challenges associated with walking to and from school. Our conclusion reflects on these results, and considers the extent to which routinized walking at the neighbourhood level challenges the social roles and subjectivities associated with automobility.

Automobility and Childhood

The restructuring of the urban fabric to accommodate and prioritize motor vehicles has transformed ways of living in, and moving through, time-space. In broad terms, it has changed everyday practices of sociality, family life, work, education and leisure, as well as increasing the distances which people need and expect to travel for everyday purposes (see both Martin and Soron, this volume; Urry 2004). It has particularly affected children, restricting their uses of public space in urban settings, reducing available play spaces and opportunities for independent travel, and increasing traffic danger.

Automobilization has contributed to a withdrawal of children from streets and public space due to their own and/or their parents' increased fear of traffic collisions (Black et al. 2001; Hillman et al. 1990; O'Brien et al. 2000; Tranter and Pawson 2001). Colin Ward was among the first to recognize that the car had 'overtaken' the city, resulting in a lack of space for children, and fewer opportunities to participate in the public life of the city:

> the assumption that the car-driver has a natural right to take his vehicle anywhere in the city has, quite apart from the threat to life, gradually attenuated many of the aspects of the city that made it an exciting and useable environment for children ... The street life of the city has been slowly whittled away to make more room for the motor car ... Whole areas which were once at the disposal of the explorer on foot are now dedicated to the motorist. The city, which used to be transparent to its young citizens who could follow the routes across it unerringly, is now opaque and impenetrable. (Ward 1978, pp. 118–21)

Withdrawal of children from urban streets is, in large part, a response to the rising traffic volume, speed and danger in Western cities that have accompanied automobility (Hillman 1993; Joshi and MacLean 1995; Mullan 2003). Large cities, in particular, are perceived to be unsafe, harsh, dangerous environments for children (O'Brien et al. 2000). Such perceptions contribute to notions of the vulnerability of child pedestrians and cyclists in urban space – and associated perceptions that good parenting requires restricting children's access to this space. This equation is reinforced by powerful social constructions of children as lacking competence to negotiate auto-dominated environments.

Indeed, critical work in the area of road safety has highlighted how public opinion, legal processes and expert analysis frequently blame the victim in child pedestrian accidents: unsatisfactory child behaviour, rather than driver error, automobile design or an unsafe environment, is commonly identified as the cause of collisions (Hillman et al. 1990; Davis and Jones 1996; Jain 2004). Such thinking – strongly evident in New Zealand (Roberts and Coggan 1994) – has supported preventative campaigns centred on improving children's skills as pedestrians. Typically based in schools, the goal of these campaigns has been to change children's behaviours, rather than the priorities of automobile-dominated environments (and the associated economic system). For Roberts and Coggan, the "strength and pervasiveness of the ideology of victim blaming in child pedestrian injuries" was explained by the economic imperative of maintaining autodependency, even "at the expense and suffering of children" (Roberts and Coggan 1994, 749).

Decreases in road casualties in most developed nations since the 1970s have been due in large part to efforts to protect drivers and other vehicle occupants. Motor vehicle industry lobbyists and government regulators have argued at length over social and legal expectations for driver behaviour, and technological changes to improve the safety of the motor vehicle for those inside it (see Wetmore's chapter). By contrast, the safety of pedestrians and cyclists has been relatively seldom commented upon. While non-motorized road users may gain some benefit from declining average speeds within urban areas, the irony of many road casualties occurring among those who are outside the automobile often merits little more than passing comment. Nevertheless, some progress in reducing pedestrian casualties has occurred in recent decades, due in part to strategies which focus on removing children from roads (Mullan 2003).

This said, the constriction of children's geographies is not uniform. In a comparison of German, British, Australian and New Zealand urban children's freedom, Tranter and Pawson (2001) found that German children had the greatest independent mobility. The authors attributed the children's mobility to higher urban densities, facilities such as schools close to children's homes, good public transport, large numbers of people on the street, and a sense of collective responsibility among German adults for children's well-being in public space.

Clearly, spatial and cultural norms vary around appropriate levels of freedom for children. Valentine observes that "most parents walk a tightrope, wavering between being anxious that they are being overprotective and fearing that they are placing their children in danger by granting them independence" (Valentine 1997, 73). Intense familial negotiations have been shown to surround the process of parents 'letting go' and allowing their children to be independent, and 'keeping them close' for protection (O'Brien et al. 2000). In Sydney, Australia, Dowling (2000) found that mothers experience pressure to chauffeur their children between activities, and that this behaviour partially constitutes social constructions of 'good parenting'. More broadly, chauffeuring is also linked to the increased scheduling of children's (and, by extension, parents') lives around privatized, institutionalized activities (Collins and Kearns 2001; Mitchell et al. 2007; Mullan 2003). The environments associated with these activities – such as indoor play spaces and childcare facilities – help to constitute contemporary children's place in the city (Karsten 2002).

Consequences of restrictions on children's mobility may include reduced knowledge of the local environment, less competence and confidence in public space, and slower emotional development (Hillman 1993; Pain 2001). Fitness can also be impaired due to a lack of opportunities to engage in routine active travel — for example, walking and cycling (Black et al. 2001; Hillman 1993; Pain

2001; SPARC 2001). Tranter and Pawson (2001) argue that continual automobile use during childhood is likely to instil car-dependent habits that will be carried over into adulthood. However, some researchers argue that increased automobile travel has not impaired children's cognitive and emotional development (Joshi et al. 1999), and others present evidence that the increased scheduling of children's lives is enjoyable for some (O'Brien et al. 2000).

In light of the foregoing discussion, it is important to acknowledge that children do not experience public space in universally similar ways. There is no "unitary public child" facing "hostile urban landscapes dominated by physical decline, threatening adults and fast-moving cars restricting their access and independent movement" (O'Brien et al. 2000, 258). Children experience significantly different levels of freedom and have different relationships with the urban public realm, depending on age, gender and ethnicity (O'Brien et al. 2000; Valentine 1997). In particular, girls may have a more restricted and narrower range than boys, and may more likely be supervised by adults in public space (Karsten 1998; 2003). The effects of gender on mobility are exacerbated for some minority ethnic groups (O'Brien et al. 2000). However, other research has revealed that boys are becoming increasingly restricted by their parents as well (Pain 2001), due in part to perceived immaturity, and a tendency to engage in boisterous and aggressive forms of play (Underwood et al. 2001).

Research has also shown that freedom to negotiate public space correlates with age. Children receive 'licences' from parents to undertake certain activities independently (such as walking to school, playing in a park, or taking public transport) at different ages, often depending on family culture, gender and ethnicity (O'Brien et al. 2000; Valentine 1997). Research in Auckland has estimated the average ages at which children are granted independence: to walk to school (8 years), cycle (10 years), and catch public transport (11–12 years) (Hall 2003). This suggests that, for all but the shortest journeys, children below age eight must be under adult supervision.

These studies illustrate the very fundamental effect that automobility has had on children's lives. While the loss of independent mobility outlined above is an obvious negative consequence, it is also important to note that, for many children, the car has extended and expanded the activities they undertake, and the social and familial contact they are able to enjoy. In this sense, the car has completely transformed experiences of childhood. Being in a car is normalized from infancy for most children in Western cities, and many aspects of the ways in which children live, learn, play, participate in activities, and experience life are fundamentally structured by the automobile. However, levels of car use among children are not uniform: one recent study in Brisbane, Australia found a strong positive relationship between household income and the likelihood of children being driven to and from school (Spallek et al. 2006). Children from the poorest households were significantly more likely to walk to school, and to walk greater distances, than those from the wealthiest homes, leading the authors to "support the generalisation that children from low SES [socio-economic status] backgrounds are exposed to greater risk of pedestrian injury" (Spallek et al. 2006, 137).

In light of the ingrained nature of car use and concerns for the vulnerability of child pedestrians, it is unsurprising that car use has come to be perceived as a way of caring for children, and as an expression of care and love for escorted family members more generally (Maxwell 2001). Indeed, Sheller suggests that cars have become 'family members' in many Western households, in the sense that they are integral to the fulfilment of social roles, and are caught up in emotional connections:

> [w]hen cars become not only devices for escaping families, but also members of families, repositories for treasured offspring and devices for demonstrating love, practising care, and performing gender, they bring into being non-conscious forms of cognition and embodied dispositions which link human and machine in a deeply emotive bond. (Sheller 2004, 232–34)

Cars are often viewed by families as facilitating greater opportunities for children, such as enabling them to attend 'better' schools, and to participate in sports, activities, or hobbies that would otherwise be relatively inaccessible by other means. However, it is also important to appreciate that while car use extends social networks and provides opportunities, automobility is primarily responsible for dispersing these networks and opportunities in the first instance, by facilitating connections over greater distances (Bean et al. 2008; see Soron's chapter).

Context

This study is situated in Auckland, New Zealand (2006 population: 1.4 million). The metropolitan area has high rates of vehicle ownership (approximately one car for every two residents) and use (approximately 78 percent of work journeys are undertaken by car) (Auckland Regional Council 2005; Laird et al. 2001). The city lacks a mass transit system, and its sprawling land use patterns have been argued to exacerbate dependency on the automobile (Auckland Regional Council 2005; Laird et al. 2001; Mees and Dodson 2001). Public transport patronage declined rapidly between the 1950s and mid-1990s, but rebounded from record low levels until 2003, thanks in part to substantial investments that improved public transport service levels. Rail use has also been increasing due to infrastructural investments such as the completion of a downtown railway station in 2003. The urban region is characterized by low-to-moderate density suburban development, within which the overwhelming majority of schools are located.

As vehicle use in Auckland has increased, concerns for safety have risen. Pedestrian injury rates have hovered at around 400 reported casualties per year for the region over the last decade (Land Transport New Zealand 2005). Pedestrian injury remains a leading cause of death for New Zealand children aged between one and fourteen (Feyer and Langley 2000), despite a widespread withdrawal of child pedestrians from the street as parents increasingly chauffeur children (Collins and Kearns 2001; Kearns and Collins 2003). In broad terms, the general public in Auckland – as elsewhere (Amato 2004) – views walking as an inconvenient, slow mode of travel, and the faster option of cycling as unacceptably risky (due to the frequency of collisions with vehicles). Specific interventions to improve the safety of child pedestrians, beyond road safety education, have been few – something which has been manifest in entrenched reluctance on the part of transport authorities to introduce reduced speed zones around schools (although a gradual, and painstaking, effort to introduce 40 km/h zones around some primary school entrances is finally underway, as of early 2007). Walking school buses (WSBs) in Auckland are not promoted primarily as pedestrian safety initiatives, although safety benefits may flow from adult supervision of children and the increased visibility that goes with walking in a group (sometimes equipped with high-visibility clothing).

Recent research into primary school children's perceptions of neighbourhood in Auckland revealed a keen awareness of traffic-related risks, which featured prominently in their photography, writing and discussions (Mitchell et al. 2007). This was most pronounced in younger children, and in those whose school was located in a particularly congested area. More generally, all the participants were aware that, because drivers are not as careful as they could be, children need to take responsibility for themselves. In one particularly telling conversation, which followed observation of a garbage truck reversing around a corner near their school, the children not only characterized the truck as a hazard they would need to manage, but added that if an accident were to happen in such a situation, the child was most likely to blame. Such comments correspond with the widespread tendency in New Zealand, noted above, to blame child pedestrians (and cyclists) for collisions with vehicles.

Notwithstanding the marginal status of walking in Auckland, and the particular dangers experienced by child pedestrians, walking school buses have enjoyed considerable success since their first trial in the region in 2000. They involve groups of children walking to and from school under adult supervision, along a set route complete with specified stops at which they may embark or disembark. Adult volunteers – almost always parents – guide the bus between its farthest stop and the school, while maintaining discipline and remaining alert for potential hazards and obstacles. The initiative has been promoted in Auckland primarily as a tool for reducing traffic congestion in the vicinity of schools, with the health benefits of routine exercise also receiving some mention. In addition, WSBs are helping to re-legitimate walking as a form of transport on suburban streets. However, as noted elsewhere (Collins and Kearns 2005; Kearns and Collins 2006) these volunteer-driven initiatives have been strongly concentrated in the most privileged neighbourhoods, where child pedestrian injury rates are typically low, and problems of childhood overweight/obesity generally least pressing.

The purpose of this chapter is not to document the strengths, weaknesses, and relative merits of WSBs. Rather, it is to explore the ways in which those parents who participate in this initiative in the Auckland region perceive the typically automobilized environments around schools, and the place of children within them. Specifically, we draw on surveys of WSB coordinators (those parents who, in addition to volunteering to walk, take on additional administrative responsibility, particularly in terms of scheduling other volunteers). Over three consecutive years (2003–05), short self-completion questionnaires were distributed to parent coordinators at all primary schools known by the Auckland Regional Transport Authority (ARTA) to be operating at least one WSB route. Respondents were asked to reflect on the successes, challenges, costs and benefits of walking over the preceding school year, and completion was tied to a small financial grant from ARTA to assist with operating expenses. The number of respondents increased steadily over this time (2003: 85 routes; 2004: 89 routes; 2005: 108 routes), reflecting increasing uptake of the initiative in Auckland, particularly within the areas of highest socio-economic status (Collins and Kearns 2005).

Results

Our 2005 survey found that 2019 children registered for WSBs in the Auckland region. This number was based on information provided for 97 routes – an average of 20.8 registered children per route. There was considerable variation between routes, from a minimum of two children to a maximum of 100. As in previous years' data, actual usage rates were considerably lower. A total of 1099 walkers were reported as walking on an average morning (89 routes reporting; mean = 12.3 children per route). The numbers for a typical afternoon were 880 children walking (70 routes reporting; mean = 12.6 children per route). The average length of a single WSB journey was reported as being twenty minutes, and most 'passengers' are aged between five and nine years (with older children 'graduating' to independent walking). Information received from respondents regarding the means by which participants would have travelled in the absence of the WSB also enabled us to estimate the number of car journeys saved as 1792 on an average day (for the 177 routes known to operate in the region at the end of 2005). For those who participate, however, routinized walking is not valued solely, or even primarily, for its contribution to reducing congestion.

In both 2004 and 2005, WSB coordinators valued most highly the benefits of health and exercise (see Table 7.1) – especially for children, but also for adult volunteers. Walking is perceived as a health promoting activity, and a response to the public health imperative of increasing physical activity.

Consistent with earlier research which found that children have internalized health promotion discourses (Kearns et al. 2003), parents coordinating WSB activities commented enthusiastically about the perceived health benefits of regular walking, with some reporting marked improvements in children's (and, sometimes, adults') fitness levels over the course of a year. Illustrative responses to questions about benefits include:

Table 7.1   Perceived Benefits of Routine Walking for School Travel

Benefit	Ranking (2005)	Ranking (2004)
Exercise/health	1st	1st
Reduced car use/congestion	2nd	4th
Sociability/sense of community*	3rd	2nd
Injury prevention	4th	3rd
Convenience	5th	7th
Teaching road safety skills	6th	–
Experiencing local environment	7th	–
Children arrive at school on time*	--	5th
Fun	--	6th

* *prom*pted response in 2004

- Creating healthy lifestyles. Enjoyment of walking. Fitness.
- The children on the bus are becoming quite fit.
- Encouraging healthy children.
- The children let off steam walking to school (which increases their fitness) [and] they are more settled upon arrival to school.

As Table 7.1 illustrates, participants are also cognizant of benefits that feature prominently in governmental discourses about adult-supervised walking to and from school: not only reduced congestion, but also injury prevention and the imparting of road safety skills. Yet another important benefit in the eyes of many respondents is sociability, and a heightened sense of community. While the contribution of schools to social cohesion has been remarked on elsewhere (Witten et al. 2003), the act of routinely walking to and from school offers an additional opportunity for social interaction, above and beyond that which may be found through conversations as the school gate, or participation in school events. In general terms, walking offers chances for conversation among strangers and casual acquaintances which are otherwise rare in an automobilized society (Bean et al. 2008). Walking school buses, by virtue of the range of parents and children involved, offer specific opportunities for social interaction between children of different ages, between parents who may not otherwise know each other, and between children and adults who are not their parents. Such interactions are valued, as the following responses to questions about the positive aspects of walking attest:

- Friendship building, children getting to know other adults they do not usually have contact with, children enjoying walking with friends, great way for new families to get to know families in neighbourhood.

- Everyone seems to enjoy the walk, the kids talk the whole way, parents like it too ... Five year olds joining the bus seem to have a healthy sense of independence when starting school ... A very positive sense of neighbourhood has been developed, the kids have enjoyed a ... Halloween trick or treat together, Christmas BBQ, end of term parties, etc., as well as playing regularly with each other. Parents have an immediate circle of people to call if needed.
- The children love meeting and walking together to school. Great friendships have been made as well as conversations enjoyed.
- [It is] a great way of bringing our neighbourhood together – [a] fantastic support network.
- [It is] a focal point in the community [that] gives neighbours a reason to communicate.
- Younger children getting to know older children at school. The older children are very supportive of the younger ones on the bus – this carries on at school.

Such sociability between families, and across different age groups, is dependent on stepping outside the confines of the private motor vehicle. Although some children and adults participating in WSBs would have walked to and from school anyway, most indicate they would have driven in the absence of this initiative. While the car is not necessarily an asocial or antisocial space, especially when being used for the transport of children, it does remain private – in the sense of normally being reserved for immediate family members. Walking, by contrast, occurs without physical barriers or strong expectations of privacy (Mitchell 2005), and at a pace that enables conversation. Walking and talking appear to go hand in hand, and contribute to the creation of socially vibrant neighbourhoods – the antithesis of socially dead 'public' spaces dominated by vehicular traffic and associated externalities.

The promotion of road safety is another commonly-reported positive experience, linked to the idea that children learn valuable practical knowledge about traffic through routine, adult-supervised walking. In the words of one respondent, "Children have learned 'awareness'. When cars come up to the school driveway, they see the car and call out 'car coming' and everyone knows to move off the driveway and stay to the left." Significantly, the perceptions of children being better equipped in terms of safety extend beyond traffic-related risk. For instance, according to one coordinator, "children are aware of safe people/houses in the community." In other words, although road safety may be foremost in some parents' minds, knowledge of the community generated through routine walking under adult supervision may also yield safety benefits.

References to learning rules for road safety point to the disciplined nature of participation in a walking school bus. The form of walking it offers is highly structured, in terms of timetable, route and expectations as to behaviour. While walking for transportation (as opposed to leisure) can be interpreted as counter-cultural, especially in auto-dominated environments such as Auckland, the specific form of walking required by a WSB also reinforces cultural notions of childhood. In particular, it contributes to perceptions of children as unpredictable pedestrians, poorly equipped for a world dominated by cars, and in need of constant adult surveillance and correction. This view is reflected not only in expectations that children will comply with adult-imposed strictures (for example, requirements that walking occurs at a steady pace, and in a more-or-less direct line), but also in expectations from schools and local government that WSBs will maintain a certain adult-child ratio (typically 1:8).

In this context, impulsive and/or playful behaviour on the part of children presents a challenge for adult volunteers. Annual surveys of WSBs are replete with complaints about children's ill-disciplined walking (e.g., walking in front of/ behind the group; on the road; too fast/too slow; failing to keep left), unreliability in meeting the bus and subversion of adult-imposed rules. Typically, the disapproved behaviours are relatively common to groups of primary school-aged children, and include pushing, fighting, and general unruliness, as well as not being fully attentive. Thus, one coordinator complained of "kids reading while walking, bouncing basketballs while walking and talking back at the driver," although we can speculate that for children themselves the first two activities (at least) are unremarkable. Responses to children's misbehaviour and ill-discipline include the reassertion of rules, invocation of school and parental authority, and – in some cases – exclusion from the group.

The reflections of adult coordinators on the challenges of walking with groups of children also invoked particular understandings of gender. With some frequency, they indicated that misbehaviour was most often associated with boys, who lacked common sense and pushed boundaries:

- Too many small/young boys makes it a difficult-to-control group.
- Keeping the boys walking rather than running/racing.
- Some boys don't want to walk because they don't want to follow the bus rules. A lot of fighting. Some boys are off the bus.
- The only negative has been a situation where we had 12 walkers, 10 of them small boys and they were a bit boisterous.

Accordingly, reported responses to behavioural challenges also had a gendered dimension:

- Talk to the boys. Reprimand them. Give them to the Principal to deal with it by informing their parents.
- Children have been put into two lines. A line for girls and a line for boys to prevent bad behaviour.

Such responses serve as a reminder of the disciplinary nature of this form of walking (Kearns and Collins 2003). This said, another WSB which had physically separated boys and girls had reportedly done so not as a result of adult concerns about children's behaviour, but at the request of the boys themselves: "At the request of the boys we usually end up with girls at the front and a 'boy bus' at the back – where they chat about Star Wars and such like! This has really helped the boys strike up some good friendships and keeps them occupied on the walk." Across our three years of survey data, girls are very seldom mentioned, and never in a negative context, suggesting that coordinators perceive them as relatively quiet and compliant walkers.

Expectations of compliant walking are not limited to child 'subjects', and complaints about the unreliability and unruliness of adults are also common. These often focus on failures on the part of parents to inform volunteers that their child will not be walking on a particular day (resulting in unnecessary delays and anxiety), as well as volunteers "not turning up for their rostered shift." Whereas in a paid employment situation, such inattention to punctuality and attendance would likely elicit censure, the voluntary basis of this walking means unreliability does not result in formal reprimand. A related negative experience concerns the recruitment and retention of volunteers, with a proportion of parents considerably more keen for their child to walk with the group than they are to volunteer their own time to assist with the operation.

While respondents' concerns about recruiting and retaining volunteers are often expressed in gender neutral terms (e.g., "we lack parental support"; "parents just aren't interested"), the overwhelming majority of those who do volunteer in WSB initiatives are mothers. Our most recent survey found that 88 percent of volunteers are women, and that a majority of routes (54 percent) have no male volunteers. In addition, 21 percent of routes have just one male volunteer, and no WSB has more men than women assisting with its operation. When asked to comment on the difficulties associated with recruiting male volunteers, 75 percent of WSB coordinators (themselves usually mothers) noted that men's work routines precluded volunteering. Comments such as "they work!", "work commitments," and "most work full-time and are therefore unable to help" were almost ubiquitous. Only a small minority of respondents (6 percent) suggested that stereotypes about gender roles were independently responsible for the lack of male involvement: generally, the notion that "most males are in full-time employment whilst mums are the main caregivers" was stated as a simple fact of suburban life.

Nonetheless, recruiting female volunteers was seldom without challenge: only 20 percent of respondents reported no problems in this area (7 percent also reported no problems in recruiting men). For 32 percent of respondents, difficulties in securing female volunteers was also attributed to 'work', while 27 percent noted that the responsibilities of caring for other family members (especially younger children of pre-school age) made it difficult for many women to be involved. While child-care responsibilities were never mentioned as factors restricting men's participation in walking, they were considered common for women, either independently or in combination with paid work. Thus, comments on the difficulties associated with recruiting women included:

- Work commitments; after school commitments; preschoolers/babies.
- [Some] mothers … work full-time; but the 'at home' mums have other commitments with other kids, i.e., babies and toddlers, or older kids needing to be taken to another school.
- [Our] route [can't] be managed with younger siblings walking or in push-chairs [i.e., strollers]. [It] has a steep hill at the beginning. [Also,] mothers of older children are returning to work.
- Work commitments, pregnancy and reluctance to commit if they have small babies.

To the extent that there is an expectation surrounding WSBs that mothers not in full-time paid employment, as opposed to parents in general, will commit to walking with children, the potential pool of volunteers is reduced. One consequence is that routes are then vulnerable to declines in volunteer support if and when mothers return to full-time work outside the home. Certainly, the data confirms that the temporal and spatial demands of work in Auckland (where long commutes, typically in sole-occupant motor vehicles, are routine) are often inconsistent with supervising children walking to and from school. As has been noted elsewhere (Beckmann 2001; Sheller and Urry 2000), automobility is associated with the intense scheduling of everyday life, including long work hours, and with the need to be highly flexible – something which is made possible by car use, and which makes the car indispensable, as Dennis and Urry argue in their chapter. For many parents, one consequence of their long commutes is unavailability for walking to and from school. Accordingly, if they want their children to be supervised on these journeys, they are reliant on the volunteer labour of other adults (if their children are to join WSBs), or dependent on alternatives such as driving children as part of their own travel to and from work.

While clearly a variety of challenges is associated with collective walking, perhaps the greatest is auto-dominated space itself. Routine walking leads volunteers to reflect critically on the hazards pedestrians must negotiate in this environment. Such reflection is evident in a variety of comments, including those focussed on driver behaviour, the built environment and the marginal status of walking/walkers in Auckland. In combination, these comments amount to a critique of auto dominance:

- [Our problem is] cars. We are just not getting through to parents, the council or the police. People are not stopping.
- There is no pedestrian crossing. [The] road is extremely busy with a 70 km/h speed limit. We eventually got council to install an island but we have 30 kids some days and two adults. It is hard to get 30 kids safely onto a small traffic island.

- It is along a very busy main arterial route with lots of side street crossing. Despite council's assurances that a crossing aide would be installed on the western side of Allendale Rd intersection with Mt Albert Rd, nothing has been done.
- A parent stands at the top of the hill with two flags and the rest of the party cross at the bottom of the hill … The parent holds the flags up high when it looks clear to cross the children. The flags are lowered when unsafe. But the cars do not slow down when they see the flags. It is still dangerous to cross because of the speed of the cars.

Traffic management issues feature prominently in accounts of negative experiences. Common concerns include "drivers accelerating too fast on to driveways," "impatient motorists when we are crossing side roads," and cars "running red lights" as well as speeding through pedestrian crossings. In addition, many coordinators have identified intersections that do not allow for safe crossing, especially by a group of children:

- One of the roads we cross (Havelock) is quite dangerous. It's very hard to see in both directions and cross safely particularly on the occasions when we have only one [volunteer].
- The Balmoral Rd pedestrian crossing is very dangerous. We have all seen cars go through red lights there. Could we get speed cameras installed to help slow traffic? The crossing is at the bottom of a hill and cars tend to pick up speed as they go down.

On some occasions, adult volunteers find themselves required to engage in risky, if selfless, behaviours:

> Council won't put a light on the crossing on Rathgar Rd. Four kids have been hit there. Once a motorbike was going so fast and wasn't going to stop but I made him and he had to skid then fell off, and used all this abusive language to me in front of the kids. It's just not good enough … We – my ladies and me – do a barrier across the road and the kids walk between us. So if anyone gets hit it's us first.

Such comments serve as a vivid reminder of the hazards associated with automobilization that have contributed to children's retreat from the street, and which pedestrians have a very limited ability to address. On the other hand, the organized and institutionalized nature of WSBs (when compared with less structured, independent walking on the part of children) also means that they can be used as leverage to achieve improvements in the local environment. In particular, councils have often been willing to undertake relatively inexpensive actions, such as cutting back overhanging vegetation, and upgrading/maintaining footpaths, in response to requests from WSB coordinators.

Conclusion

Automobility has led not only to the retreat of children from public space in many Western cities, but to a more fundamental restructuring of children's socialization and experiences. While this restructuring should not be considered entirely negative, it is not without a price. Typically, this includes a significant risk of injury for child pedestrians and cyclists, and diminished opportunities to explore urban spaces, especially independent of adult supervision. Children themselves are often aware of the ways in which their lives are constrained by fear, and many savour opportunities to explore outdoor environments when these are presented (Mitchell et al. 2007). Given the relatively short distances between home and primary school for many suburban children, daily travel between these two sites offers an important opportunity for routine active travel, and exploration of the local environment.

WSBs encourage and facilitate walking to and from school. These initiatives are necessarily limited in their ability to effect change in an auto-dominated city, but are sometimes portrayed as an affordable panacea for urban ills ranging from congestion to childhood obesity. Actual benefits appear rather more modest. In broad terms, the act of supervised walking challenges only some of the social practices associated with automobilization. For example, respondents are insistent that children do have a legitimate place on the sidewalks of the city; they are entitled to walk to and from school (albeit in a regulated manner). Obstacles to the realization of this entitlement – ranging from dangerous driving, to intersections which cannot be safely navigated, to deteriorating pedestrian infrastructure – are portrayed as unacceptable. In this way, coordinators' arguments do contest ingrained notions that children must 'yield' to cars.

The involvement of relatively large numbers of adults in walking with groups of children also suggests an emergent understanding of walking as an appropriate means of caring for children. While, in some circumstances, car use may be a measure of parental (especially maternal) contribution to children's well-being (Dowling 2000), many parents in Auckland have come to see the supervision of children's walking in a similar light. Their generally positive views of walking are underpinned by a range of perceived benefits, including health and fitness, and the sociability and sense of community that is fostered by regular conversations among relatively diverse groups of children and adults.

Ultimately, however, children undertaking a supervised walk to and from school do not threaten to revolutionize the social practices associated with automobility. Many parents, including the majority of fathers, are apparently unable to volunteer to walk with children due to demands of full-time work. Typically, they are commuting at the time that children depart to school – traversing the physical separation between home and work that has expanded greatly with the rise of the automobile. In addition, understandings of adult supervision and control are inherent in the WSB concept, signalling enduring concerns about the (in)ability of children to negotiate automobilized space independently. In light of such issues, it is possible to argue that supervised walking reinforces some of the social identities associated with automobile-dominated space (that is, those around gender, and the public incompetence of children). This said, it is clear that routine walking in groups also helps to break down the social isolation often associated with automobilized family lifestyles, and alerts adult participants, in particular, to some of the costs and dangers associated with allowing motor vehicles to dominate the public spaces of everyday life.

References

Amato, Joseph Anthony. 2004. *On foot: A history of walking*. New York: New York University Press.

Auckland Regional Council. 2005. *Moving forward: Auckland regional land transport strategy 2005*. Auckland: Auckland Regional Council.

Bean, Catherine, Kearns, Robin A. and Collins, Damian C.A. 2008. Exploring social mobilities: Narratives of walking and driving in Auckland, New Zealand. *Urban Studies* 45(13): 2829–48.

Beckmann, Jörg. 2001. Automobility – A social problem and theoretical concept. *Environment and Planning D-Society and Space* 19(5): 593–607.

Black, Colin, Collins, Allan and Snell, Martin. 2001. Encouraging walking: The case of journey-to-school trips in compact urban areas. *Urban Studies* 38(7): 1121–41.

Collins, Damian C.A. and Kearns, Robin A. 2001. The safe journeys of an enterprising school: Negotiating landscapes of opportunity and risk. *Health and Place* 7(4): 293–306.

——. 2005. Geographies of inequality: Child pedestrian injury and walking school buses in Auckland, New Zealand. *Social Science and Medicine* 60: 61–69.

Davis, Adrian and Jones, Linda J. 1996. Children in the urban environment: An issue for the new public health agenda. *Health and Place* 2(2): 107–13.

Dowling, Robyn. 2000. Cultures of mothering and car use in suburban Sydney: A preliminary investigation. *Geoforum* 31: 345–53.

Feyer, Anne-Marrie and Langley, John D. 2000. Unintentional injury in New Zealand: Priorities and future directions. *Journal of Safety Research* 31(3): 109–34.

Hall, Debra. 2003. *The trip to education: A market information study for infrastructure Auckland and the Auckland Regional Council*. Auckland: Research Solutions.

Hillman, Mayer, ed. 1993. *Children, transport and the quality of life*. London: Policy Studies Institute.

Hillman, Mayer, Adams, John and Whitelegg, John. 1990. *One false move: A study of children's independent mobility*. London: Policy Studies Institute.

Jain, Sarah S. Lochlann. 2004. 'Dangerous instrumentality': The bystander as subject in automobility. *Cultural Anthropology* 19(1): 61–94.

Joshi, Mary Sissons and MacLean, Morag. 1995. Parental attitudes to children's journeys to school. *World Transport Policy and Practice* 1(4): 29–36.

Joshi, Mary Sissons, MacLean, Morag and Carter, Wakefield. 1999. Children's journey to school: Spatial skills, knowledge and perceptions of the environment. *British Journal of Developmental Psychology* 17: 125–39.

Karsten, Lia. 1998. Growing up in Amsterdam: Differentiation and segregation in children's daily lives. *Urban Studies* 35(3): 565–81.

———. 2002. Mapping childhood in Amsterdam: The spatial and social construction of children's domains in the city. *Tijdschrift voor Economische en Sociale Geografie* 93(3) August: 231–41.

———. 2003. Children's use of public space: The gendered world of the playground. *Childhood* 10(4): 457–73.

Kearns, Robin A. and Collins, Damian C.A. 2003. Crossing roads, crossing boundaries: Empowerment and participation in a child pedestrian safety initiative. *Space and Polity* 7(2): 193–212.

———. 2006. Children in the intensifying city – Lessons from Auckland's walking school buses. In *Creating child-friendly cities: Reinstating kids in the city,* Brenda Gleeson and Neil Sipe, eds, pp. 105–20. London: Routledge.

Kearns, Robin A., Collins, Damian C.A. and Neuwelt, Patricia M. 2003. The walking school bus: Extending children's geographies? *Area* 35(3): 285–92.

Laird, Philip, Newman, Peter, Bachels, Mark and Kenworthy, Jeffrey R. 2001. *Back on track: Rethinking transport policy in Australia and New Zealand.* Sydney: University of New South Wales Press.

Land Transport New Zealand. 2005. *Road safety issues.* Auckland: Land Transport New Zealand.

Maxwell, Simon. 2001. Negotiations of car use in everyday life. In *Car cultures,* Daniel Miller, ed., pp. 203–22. Oxford: Berg.

Mees, Paul and Dodson, Jago. 2001. The American heresy: Half a century of transport planning in Auckland. In *Geography: A spatial odyssey: Proceedings of the third joint conference of the New Zealand Geographical Society and the Institute of Australian Geographers,* Peter G. Holland, Fiona Stephenson and Alexander E. Wearing, eds, pp. 279–87. Hamilton: Brebner Print.

Mitchell, Don. 2005. The SUV model of citizenship: Floating bubbles, buffer zones and the rise of the 'purely atomic' individual. *Political Geography* 24(1) January: 77–100.

Mitchell, Hannah, Kearns, Robin A. and Collins, Damian C.A. 2007. Nuances of neighbourhood: Children's perceptions of the space between home and school in Auckland, New Zealand. *Geoforum* 38(4) July: 614–27.

Mullan, Elaine. 2003. Do you think that your local area is a good place for young people to grow up? The effects of traffic and car parking on young people's views. *Health and Place* 9: 351–60.

O'Brien, Margaret, Jones, Deborah, Sloan, David and Rustin, Michael. 2000. Children's independent spatial mobility in the urban public realm. *Childhood* 7(3): 257–77.

Pain, Rachel. 2001. Gender, race, age and fear in the city. *Urban Studies* 38(5–6): 899–13.

Roberts, Ian and Coggan, Carolyn, 1994. Blaming children for child pedestrian injuries. *Social Science and Medicine* 38(5): 749–53.

Sheller, Mimi. 2004. Automotive emotions: Feeling the car. *Theory Culture and Society* 21(4/5): 221–42.

Sheller, Mimi and Urry, John. 2000. The city and the car. *International Journal of Urban and Regional Research* 24(4): 737–57.

Spallek, Melanie, Turner, Catherine, Spinks, Anneliese, Bain, Chris and McClure, Rod. 2006. Walking to school: Distribution by age, sex and socio-economic status. *Health Promotion Journal of Australia* 17(2): 134–38.

SPARC. 2001. *SPARC facts 97–0: Sport and recreation New Zealand.* Wellington: SPARC.

Tranter, Paul and Pawson, Eric. 2001. Children's access to local environments: A case study of Christchurch, New Zealand. *Local Environment* 6(1): 27–48.

Underwood, Marion K., Galen, Britt R. and Paquette, Julie A. 2001. Top ten challenges for understanding gender and aggression in children: Why can't we all just get along? *Social Development* 20(2): 248–66.

Urry, John. 2004. The 'system' of automobility. *Theory Culture and Society* 21(4/5): 25–39.

Valentine, Gill 1997. "Oh yes I can." "Oh no you can't": Children and parents' understandings of kids' competence to negotiate public space safely. *Antipode* 29(1): 65–89.

Ward, Colin. 1978. *The child in the city.* London: Architectural Press.

Witten, Karen, McCreanor, Tim and Kearns, Robin. 2003. The place of neighbourhood in social cohesion: Insights from Massey, West Auckland. *Urban Policy and Research* 21(4): 321–38.

PART 3

Inevitable Automobility?

Chapter 8

The Politics of Mobility: De-essentializing Automobility and Contesting Urban Space

Jason Henderson

Political contestation of automobility is unfolding around the world, as noted in Martin's chapter. New discourses and practices that temper automobility are attempting to reconfigure urban space into a development pattern broadly labelled 'smart growth' (a development strategy based on increased density, transit corridors and regional coordination) and 'new urbanism' (an architectural concept focused on mixed use, walkable neighbourhoods) in North America, and 'compact cities' in Europe and globally. These new movements have the potential to lead to what Dennis and Urry (this volume) call a 'post car future' tipped into a different path from the privatized petroleum-steel automobility system.

Yet, because of the self-expanding character of the car system, Dennis and Urry suggest, the automobility system is impossible to undo and the spatial configurations facilitating public transit are irreversibly lost. In a compelling look into how Chilean neoliberalism has adopted automobility as an important element in its expanding agenda, Trumper and Tomic's chapter suggests that any confrontation with automobility may require changing neoliberal consumerist, individualistic values, which have become globally intractable. In this light automobility appears as a given, making political challenges to automobility seem futile, and overlooking how it is a site of struggle. For many the automobile has been essentialized.

My case study of the politics of mobility in Atlanta, Georgia, illustrates how local-scale contestation of hyperautomobility can provide insights into the complexity of discourses and practices related to motorization resistance and post-car mobilities. This analysis of contestation calls into question essentialized discourses about automobility and provides an opening to counter automobile hegemony. In particular I examine how some neoliberal capitalist interests may oppose excessive automobility and engage in local political debates to temper the car, yet centre their opposition on an essentialization of automobility.

In what follows, I focus on the complicated outcome of a local coalition to produce alternative transportation to the automobile. In drawing on my analysis of media discourses, participant observation, and interviews with stakeholders, during heated debates about public transit, I argue that it is not possible to understand Atlanta's 'love affair' with the automobile without considering how it directly serves a secessionist ideology and practice. This 'secessionist automobility' is about using the car as a means of physically separating oneself from spatial configurations that are thought to be threatening or not aligned with particular values and ideas about the city and space. The automobile is an instrument of spatial secession enabling a particular normative vision of the city to be achieved. This chapter reveals nuances of secessionist automobility, in which the automobile enables not only racially motivated whites, but also others with an anti-urban ethos to move to Atlanta's outer suburban areas or navigate the city without having to interact with others.

De-essentializing Automobility

Vuchic (1999) laments that academics and policymakers have adopted an 'inevitability hypothesis' in automobility discourse, which suggests that present trends in the growth of automobility are natural and inevitable. Policymakers and others have essentialized the automobile. Such influential scholars and prestigious research bodies as the Transportation Research Board (TRB) in the United States have adopted the inevitability hypothesis. Indeed, the TRB (2001), which provides advice to the US Congress on transport matters, concluded in 2001 that, if it means limiting parking supply, increasing fuel taxes, or taking away road space, American politicians are not interested in making cities more transit-friendly and less automobile-dependent. Unable to seriously consider alternatives that contest the spaces of automobility, the TRB emphasizes 'clean' fuels and hybrid cars, congestion management through pricing, and 'intelligent' transportation systems.

Similarly, in the UK a 'predict and provide' policy towards automobiles dominates transportation discourses despite calls for sustainable transport investments and to implement policies to reduce automobility (Vigar 2002; Docherty 2003). Meanwhile, much of the developing world, the media claims, is having a 'love affair' with cars (Luard 2003; Sperling and Claussen 2004). And as China rapidly motorizes, some cities have restricted bicycle use (Pucher et al, 2007; Martin, this volume).

The idea of a love affair arises in part from the claim that the automobile embodies such values as individualism, freedom and democracy so dearly held in Western societies, especially the United States (Dunn 1998). This view, however, contradicts the reality that, as an established and institutionalized system, automobility coerces individuals into driving. Automobility subordinates all other modes of transport and ways of dwelling, requires enormous state subsidy and regimentation of urban space for maximum through-put and speed, and requires a centralized state-backed capitalist oligopoly of oil, highway, automotive manufacturing and real estate control over transportation policy (Freund and Martin 1993; Urry 2004; see Soron's chapter). As a result, transportation policy-makers resist policies that restrict or limit automobility out of fear of upsetting an electorate they presume is universally 'in love' with automobility.

The central task of this chapter is to deconstruct the universalization of automobility and to make a distinction between the cultural or emotional essentialization of automobility on the one hand, and the power emanating from the automobile industrial complex on the other. Rhetoric about the 'love affair' has severely limited serious efforts to create an ecologically sound and socially just urban future. This essentialization of automobility averts and prevents a fuller understanding of its cultural, political, social, technical, and economic contexts.

If we are to expand our politics of possibilities about urban futures, we must de-essentialize automobility. Neither the auto-industrial complex nor a love affair are the only forces shaping state policy. One way of de-essentializing automobility is by probing deeper into the discourse and motivations of stakeholders in debates over automobility and urban space and to consider how mobility is not just movement but also an extension of ideologies and normative values about how the city should be configured and by whom. Just as Lefebvre (1991) theorized that the character and nature of produced space reflects dominant modes of production and social relations (and power) within a given society, we must consider how forms of mobility contain embedded social relations. In this chapter I explore how automobility reflects a complex expression of values that are then manifested in local political conflicts shaping urban spaces and social relations.

More specifically, the task of de-essentializing automobility can be furthered by case studies of sites of contestation over automobility and urban space. This chapter offers such a case study by examining the politics of automobility in Atlanta, Georgia, a post-industrial, information-oriented city with strong ties to the global corporate system. When the federal government in 1999 targeted Atlanta as one of the worst air polluters in the US, as a punitive measure it suspended allocations which made up 80 percent of funding for most major road projects in Atlanta the city. As a result, Atlanta became a national focal point in debates over automobility. The federal suspension of road money was unprecedented – it had never before punished a city for having too much automobility.

Contesting Automobility in Atlanta

Atlanta is one of the fastest growing metropolitan areas in the US. Between 1990 and 2000, its population increased over 40 percent, from 2.9 million to 4.1 million. Expectations are that by 2025, 2.3 million more people will live in Atlanta (ARC 2004b). During the 1990s, Atlanta rose as a leading post-industrial job growth centre, and specifically, as a node in the global telecommunications and air travel network. Like most North American cities, and increasingly parts of some global cities, everyday life in Atlanta centres on the automobile. Atlanta is the embodiment of hyperautomobility as described in Martin's chapter: of saturated car ownership, and high rates of daily car trips over long distances, with few car occupants. The average person in Atlanta drives 30.5 miles a day (ARC 2004b). The average Atlanta commuter spends 67 hours, or over 8 working days a year, in congested conditions. Because of congestion that lasts 8 hours a day, the average commuter consumes 46 excess gallons of gasoline annually, and loses $1,127 a year in equivalent lost time (Texas Transportation Institute 2004).

Faced with negative national press focusing on Atlanta's congestion and smog problems, Atlanta's corporate elite 'growth machine' – a handful of Fortune 500 companies, real estate investment trusts, and one of the nation's largest energy companies – led in the creation of a new state agency, the Georgia Regional Transportation Authority (GRTA, pronounced 'Greta') (Henderson 2004). This unique agency was to force parochial local officials to clean Atlanta's air by thinking regionally about transportation planning and getting people to reduce their driving. Public commentators thought the corporate-dominated GRTA heralded a new progressive political era in Atlanta, with the potential to reshape mobility and urban space. An emerging coalition of corporate elites, civil rights leaders, and environmental and neighbourhood activists envisioned GRTA as the conduit for expanding public transit and implementing land use regulations designed to reduce automobility.

In critically questioning automobility, the coalition dramatically transformed local transportation politics. The multiple factions collectively promoted a mobility vision through grassroots advocacy, publicity campaigns, direct lobbying of public officials, and engagement in the electoral process. In framing their mobility vision of reducing automobility and reconfiguring urban space, many urban activists in Atlanta articulated environmental ethics or social justice concerns. Civil rights and environmental justice advocates outlined a mobility vision based on the right of access to the city and its amenities. Corporate elites articulated mobility visions such as support for recentralization of development and enhancement of rail systems to bolster property values and the exchange value of the entire region in order to promote an ideology of growth.

After decades of hegemonic control over Atlanta's transport and development policies, Atlanta's local highway-industrial complex (builders, suppliers, and services for automobility), was put on the defensive. This faction in Atlanta's politics of mobility was represented by the Georgia Department of Transportation (GADOT) and the Georgia Highway Contractors Association (GHCA), a collection of construction companies, and some real estate interests.[1] The epitome of this camp was the 'Gwinnett Mafia' (McCosh and Shelton 1999a, 1999b, 1999c). These were landed interests centred in (but not limited to) suburban Gwinnett County who planted key allies in all levels of transportation decision-making in Georgia, including the powerful GADOT and the regional planning agency – the Atlanta Regional Commission (ARC). From these venues they actively promoted a massive highway expansion program for metropolitan Atlanta, and were complemented by a local non-profit highway advocacy organization called Georgians for Better Transportation (which also had local construction, automobile and oil firms as members). Meanwhile, national industry trade organizations,

---

1   For a more extensive discussion of this faction, see Henderson 2002.

such as the American Road and Transportation Builders Association (ARTBA) and the American Highway Users Alliance (AHUA) became actively involved, identifying Atlanta as a key site of the national sprawl debate. With vested interests in automobility, these local and national organizations articulated a vision of more roads and parking space, but in a new defensive posture they also articulated limited transit expansion in congested corridors.

Meanwhile, just as Atlanta's corporate elite were confronting the environmental and social problems of automobility and assuring investors that they were capable of solving them, resistance from another quarter further complicated the emerging contestation of automobility. In 2000, a controversy, encapsulating racial, gender and class-based rhetoric about public transportation, hit Atlanta. The racial crisis centred on John Rocker, a young white male player for the local professional baseball team, the Atlanta Braves, who had delivered a racially charged homophobic diatribe to the national sports media. It began when his bigoted comments to a reporter about New York's subway were widely published:

> Imagine having to take the number seven train to the ballpark, looking like you're riding through Beirut next to some kid with purple hair next to some queer with AIDS next to some dude who just got out of jail for the fourth time next to some 20-year old mom with four kids. It's depressing. (John Rocker, quoted by Pearlman, 1999, N.p.)

Reflecting on the racist diatribe, syndicated columnists and social commentators asked "What does it say about us?" (Schneider 2000). Civil Rights historian David Garrow (2000) called Rocker a 'human Confederate Flag'. Garrow also stressed that the intolerance expressed in the diatribe was shared by many whites, and white fans' cheers early in the next season compelled some locals to ask if the region's 'redneck underbelly' had been exposed (Smith 2000). The commentary proved embarrassing enough to Atlanta's corporate elite that they called upon old stalwarts of Atlanta's Corporate/Civil Rights regime (Stone 1989) such as former Atlanta mayor Andrew Young and retired baseball great Hank Aaron to defuse the situation and beg for renewed racial healing (Young 2000).

While media reports focused on the intolerance and racism in Rocker's remarks, they missed the spatial context of automobility and vitriolic hostility towards transit and diversity of urban life, exemplified in the reference to experimental youth, homosexuality, teenage mothers and non-Western countries. Given Atlanta's hyperautomobility, it is not too surprising that, during the bigoted rant, John Rocker was speeding down a massive multi-lane freeway in a large SUV (a Chevy Tahoe). In venting to the reporter his disdain for New York's subway, he yelled obscenities and gestured at other motorists from within his speeding cocoon. Continuing to speed while holding the steering wheel in one hand and a cell phone in the other, he said that the thing he hated more than anything else in the world was traffic.

I have no patience. So many dumb asses don't know how to drive in this town. They turn from the wrong lane. They go 20 miles per hour. It makes me want – Look! Look at this idiot! I guarantee you she's a Japanese woman. How bad are Asian women at driving? (John Rocker, quoted by Pearlman, 1999, N.p.)

The woman was white, but to this angry white male, everyone else on the road was in his way (reflecting the egoism noted by Conley in this volume), driving too slowly or in the wrong lane, or not signalling properly. Traffic, women drivers and minorities were oppressing him. He did not consider how his driving was part of the problem. Rocker was disdainful of possible alternatives to his SUV and had no patience with an alternative way of life – for example a vision of smart growth or new urbanism – based on a compact urban form with intensive transit infrastructure containing pedestrian and transit spaces, where people would have physical proximity to 'others' of different racial, class, gender or sexual orientation. His SUV was more than just an instrument for traveling through the city. It was an instrument of secession from what he scorned in contemporary American urban space. Public transit was a warren for 'AIDS and welfare queens'. Times Square, a high density public space shared by pedestrians, buses, taxis and cars, was full of "too many foreigners who don't speak English" (John Rocker, quoted by Pearlman, 1999, N.p.). Trading in the SUV for a transit pass, and the house on an acre lot in segregated, low-density suburb for denser, mixed-use developments with shared public spaces was the antithesis of his values and ideologies about space and how he preferred to live.

Unwittingly, in using the car as a means of physically separating himself from spatial configurations of higher urban density, public space, or the city altogether, this angry white male baseball star was practicing a distinctive politics of secessionist automobility. Secessionist automobility is engendered by what Vuchic (1999) characterized as the relatively low out-of-pocket expense of automobility for the middle and upper classes trading off distance for other costs. Households that seek to distance themselves from poor schools, urban crime, racial diversity, or any other perceived urban blight are able, because of automobility, to secede from spaces where these 'problems' exist. Though, as I show later, not all secessionist automobility pivots on racism, the racism of secessionist automobility may be cautious and circumspect, and it can bring with it a blunt racialized, anti-urban politics, as Rocker and his fans exhibited. Whether circumspect or blunt, it can have a profound impact on how urban space is produced.

None of the mainstream press articles that ensued after Rocker's diatribe made the connection between automobility and space. Despite concerns about smog and the federal government's suspension of transportation funds, Atlanta's public discourse essentialized automobility. Automobility was universal and had nothing to do with racism. But significantly, in this case, the Atlantan attachment to the automobile was not so much driven by its 'love affair' with automobiles, or even with movement. Rather, it was driven by its reliance upon the automobile for achieving secession.

Rocker's rant against the diverse space of public transit resonated with decades of vitriolic anti-transit rhetoric in Atlanta (and arguably, in cities throughout the United States). This racialized animosity towards transit rests on and helps to produce automobile dependency for Atlantans who can afford it. As exhibited in Figure 8.1, Atlanta has limited public transit. Since its establishment in the 1960s, a secessionist and racist logic led to many referring jokingly to the Metropolitan Atlanta Rapid Transit Authority (MARTA) as 'Moving Africans Rapidly through Atlanta'. While thousands of white families relocated from the city centre to the suburbs in racialized reactions to the Civil Rights movement, every county in metropolitan Atlanta (except Fulton and DeKalb) had contentious local debates on joining MARTA or establishing an independent, stand-alone transit system. Gwinnett County, to the Northeast of downtown Atlanta, had its first county-wide referendum on joining MARTA in 1971, a second in the 1980s, and a third in 1990 (Cordell 1987; *Atlanta Journal-Constitution* 1998; Torpy 1999). All three failed under a cloud of racialized rhetoric, and considerable movements of middle

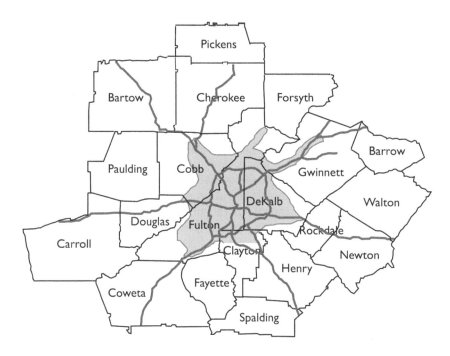

Figure 8.1   The Transit Détente in Atlanta has limited public transit, in the urban core and the skeletal system radiating outward. Based on ARC 2004a, 2004b

Source: Based on ARC 2004a; 2004b.

class whites away from proximity to blacks and to separate majority white suburbs (Keating 2001). "The reason is 90 percent racial" proclaimed the MARTA board chair in the 1980s (Cordell 1987, A1). For these whites, automobility enabled both physical secession to outer suburban areas and a means of travel through spaces inhabited by blacks, without having to interact with them.

For several decades, coverage of transit debates in the *Atlanta Journal-Constitution* revealed how deeply race mattered. In suburban Cobb County, the chairman of a local anti-tax organization declared that "MARTA-style mass transit would lead to an increase in crime and the construction of low-income housing in Cobb County" (*Atlanta Journal-Constitution* 1998, A10). Even though it was controlled by whites from Atlanta's corporate elite, racists reviled MARTA as a black-controlled urban agency in a black-run city with a black majority population.

On the heels of the controversy over Rocker's very public diatribe, the Georgia Association of Highway Contractors ran television spots in 2001 reacting to the suspension of federal road money. The accompanying video footage showed grim apartment blocks and black people getting off a bus (Ward 2001). Although the business-led GRTA publicly promoted transit, the ad warned that 'radical environmentalists' threatened to take away Atlantans' right to drive and live where they want. In another high-profile instance, CNN reported that a couple in the exurban sprawl north of Atlanta stated that they moved to the county because they felt mass transit would never come there and "transit makes areas accessible for lower-income families that could otherwise not come out here because they don't have transportation and that's good" (Wood 2000, N.p.).

The interviews I conducted with local stakeholders in Atlanta's politics of automobility corroborated this media discourse.[2] Most interviewees confirmed that white racism complicated decision-making about transit. Suburban elected officials acknowledged that a substantial proportion of their constituents held racist views. One county official said that at public meetings in her Atlanta suburb, residents loudly protested MARTA bus service because blacks would steal TVs (Interview one 2001). Another prominent suburban politician recalled that when a park-and-ride lot opened in his district to provide transit to the 1996 Olympic Games phone lines in county offices were overwhelmed by racialized, anti-transit anger (Interview two 2001). Several interviewees confided that while they were not personally racist, they understood that many of their constituents were. To remain in office, politicians had to 'represent' their constituents by resisting transit expansion or higher density development that sought to reduce automobility. They ensured that spaces of automobility would be the default built form of Atlanta.

---

2   Fifty interviews were conducted with key stakeholders in Atlanta's transportation and sprawl debate in 2001.

Secessionist Automobility and the Production of Anti-Urban Space

Racism has much to do with secessionist automobility. The racially motivated physical movement of whites to outer suburban areas in American cities is enabled by automobility, and automobility also enables travel through spaces inhabited by blacks or other minorities without having to interact with them. Moreover, the politics of race has limited the geography of transit, forcing automobile dependency by design. But secessionist automobility is not simply racially motivated. Interviewees for this research were emphatic in distinguishing racism from an anti-urban ethos, revealing nuances in secessionist automobility. Automobility was a device to achieve a spatial vision of rural ideals attached to an anti-urban image of the city as a place of vice and immorality.

For example, in public meetings focused on establishing higher density, mixed-use, and walkable 'village centres' in a fast-growing suburb, one planner noted in exasperation that the whole idea was criticized and watered down by citizens who associated the term 'village' with liberal, big government politics, and wanted nothing to do with that (Patton 1998). This ethos equated compact, new urbanist development with negative connotations, and transit with 'big city problems' like graft.

What prompts this anti-urban thread of secessionism? Certain conceptualizations of family and religion may play a role. Goldfield (1982) suggests that emphasis on personal responsibility towards one's family results in a lack of civic or social responsibility towards public space or notions of community. In contemporary American political rhetoric, 'personal responsibility' towards one's family can translate into lack of interest in collectively solving larger-scale problems such as congestion, pollution, or inequality. Instead, it is 'responsible' to move the family away from these problems – to secede. Meanwhile, Reed (2001) argues, an extreme evangelical religious worldview in some households translates into a strong anti-urban rhetoric. The religious ethos holds a pessimistic view of human nature, and therefore people, especially strangers, are not to be trusted. On this view, a dense city, which by definition has more strangers, amplifies fears about the possibility of vice. Automobility enables one to circumvent, if not secede from, the perceived evils of the city.

In this combined vision of rural idealism, 'family values' and fundamentalist religion, the low-density suburbs and exurbs of America surround corrupt cities of ghettos, vice, and mob rule (Beauregard 1993). The 'community' where these anti-urban values are synthesized moves inside, to the exclusionary private spaces (which exclude the undesired) of homes, churches, clubs, and cars. Everyday interaction with other people is homogeneous, with church and family comprising community instead of a broader multicultural, ethnic, or religiously diverse community. Private consumption of the home and by the family, or what Harvey (1989, invoking McPherson 1962) described as 'possessive individualism' takes precedence over public consumption. This view prefers private yards and private malls over public parks and civic spaces, and most importantly for the purpose of this chapter, private automobiles over public transport. As McLean (this volume) also notes, Mitchell (2005) calls this the 'SUV model of citizenship' centred on privatized, unhindered, cocooned movement through public space, whereby people feel they have a right not to be burdened by interaction with anyone or anything they wish to avoid. As in Chile, where it allows the middle class to avoid the poor, and promotes an individualist, consumerist ethos (see Trumper and Tomic's chapter), but in a different ideological framework, automobility in Atlanta is for many a vehicle for spatial secession.

The physical manifestation of this secessionist automobility vision is what Goldfield (1997) called 'countrified cities'. Low-density, single-detached houses, accessible only by automobiles, allow for proximity to rural ideals and nature. Everyone drives everywhere for everything. Yet because metropolitan areas contain the jobs and other urban services upon which modern life depends, this vision cannot be met in a practical sense, and secession is incomplete. Secessionists still must live in reasonable driving range of the city, and thus opt for the low-density subdivision within, ideally, a 30-minute commute by car to work and shopping.[3] It is for this reason that the average Atlantan drives (or is driven) over 30 miles per day and that Atlanta's sprawling 'commutershed' extends over 100 miles from the centre in every direction.

Rooted in both anti-urbanism and racism, the spaces of secessionist automobility conflicted with the spatial visions advocated by many of Atlanta's corporate elites, environmentalists and civil rights advocates, who collectively promoted a vision to redirect Atlanta's development back towards the centre, in compact patterns around transit and walking. The spaces of secessionist automobility also conflicted with the spatial vision of Atlanta's local highway-industrial complex which sought to create a new outer beltway. The latter should not be confused with the 'downtown-centred growth machine' (Logan and Molotch 1987) embodied by the corporate elite that created the Georgia Regional Transportation Authority (GRTA). This growth machine is a coalition of Atlanta-based corporate, utility, and real estate interests seeking to re-shape transportation policy in Atlanta in a way that ensures Atlanta's economic dominance in the Southeast (Henderson 2004). Notably, as the mobility crisis loomed the downtown centred Atlanta Chamber of Commerce added 'Metro' to its name and promoted itself as a regional chamber, representing more than just businesses in the city of Atlanta. These diverse groups – including environmental and civil rights groups, and the local highway-industrial complex – held specific and competing mobility visions and played an important role in the debates (Henderson 2004). Here, however, I focus on the political interaction between Atlanta's capitalist downtown-centred growth machine and secessionists.

---

3   See Gordon et al. 1991 for an apologetic discussion of the relationships between commute travel times and preferences for lower density.

Transit Détente

The space that the contestation of secessionist automobility in Atlanta produced can be characterized as a political 'transit détente'. It arose out of a struggle between competing visions of how space should be organized and for whom. It resulted in the status quo of continued automobility with limited transit, characterized by mediocre rail and bus service in the urban core and a skeletal express bus system radiating outward, with rail stations and bus stops surrounded by 33,000 park-and-ride spaces. Hence a détente, a temporary relaxation and settlement, rather than an ambitious proposal to expand transit, was chosen by many local elites. As suggested in Figure 8.1, by 2025, only 10 percent of work trips will be on transit, and only 40 percent of the population will live within a half-mile of transit (ARC 2003). Roadway expansions in outer suburbs will be numerous, and almost 80 percent of commuters will continue to drive alone to work. Despite the fact that the recent debates gave attention to automobility's negative consequences, comprehensive solutions to its problems remain elusive.

As a produced space, the transit détente is the outcome of a struggle arbitrated under capitalism, with capitalists mediating secessionist automobility. Secessionist impulses conflict with capitalist imperatives that seek to preserve the exchange value of the metropolitan region by ensuring that air pollution, congestion, and other problems of hyperautomobility do not devalue the entire area. While Atlanta's growth machine was aligning with environmentalists and civil rights advocates to promote compact, walkable, transit spaces, with GRTA as a conduit, secessionists were actively fighting transit and increased density, and thousands cheered John Rocker's angry diatribe. How did the capitalist growth machine arrive at the transit détente with secessionists? Two examples show how this détente came about.

First, Atlanta's corporate elite's creation of the Georgia Regional Transportation Authority (GRTA) – in response to the federal government's suspension of road funds due to air pollution problems stemming from automobility – tempered secessionist opposition to transit. In return for lifting the federally mandated suspension, GRTA requires that any county with a smog problem must accept public transit in exchange for receiving road funds. In acting as a referee, GRTA ensures that all localities commit to the greater goal of keeping Atlanta competitive in the global economy. If a local county or city in the metropolitan region does not show a commitment to reducing its share of smog, the authority has the power to restrict road funds and redirect them elsewhere. Hence some of Atlanta's more vehemently anti-transit counties now have, or plan to have, some sort of limited bus service (Long 2001).

In addition, GRTA's insistence on extending transit into Atlanta's sprawling suburbs serves capitalist demands for access to labour. Secessionist land use policies, such as exclusionary zoning (restricting proliferation of apartments or lower priced housing), have exacerbated 'spatial mismatch' (Ihlandfeldt and Sjoquist 1998; Nelson 2001). Since many low skill, low wage jobs in the retail and service sector are located in the suburbs, they are often vast distances from the central city and inner suburbs where low wage, low skilled workers live. In response, and as part of the mission of GRTA, Atlanta's corporate interests publicly promoted bus transit in select corridors to enable low skilled workers to access far-flung suburban jobs.

Using GRTA's logo rather than the stigmatized MARTA logo has helped to mitigate the long standing white, racialized resistance to transit. "There is too much racism in Cherokee for transit in any real sense, but maybe express buses (run by GRTA) would be acceptable," said one suburban politician (Interview one 2001). In some suburban counties retail and service firms even unified to tax themselves to jump start limited bus service and avoid controversial public referendums (Shelton 1998). "Mobility 2030," Atlanta's newest iteration of long range transportation planning, proposed express buses on radial routes from central Atlanta to major suburban activity centres which needed low-wage retail workers (ARC 2004b). Again, the overarching demands of capitalists, in this case access to cheap labour, superseded, albeit in a limited way, the spaces of secessionist automobility. In this sense, the demands of capital pre-empted local secessionist tendencies.

A second example of a transit détente involved a tacit concurrence between secessionists, the growth machine, and the environmental and social justice movements based in the urban core. In this case, all sides opposed a proposed 59-mile 'Northern Arc' freeway (see Figure 8.1), however, for differing reasons. The Northern Arc was proposed by the automobility growth machine and championed by the Georgia Department of Transportation and development interests on Atlanta's northern outer suburban fringe (Henderson 2004). With strong anti-transit ideologies, and fully ensconced in a lifestyle centred on automobility, one might expect that secessionists would welcome more highways. But across Atlanta's northern exurban fringe, secessionist opponents of the proposed road deployed a rhetoric of defending 'small town quality of life' (Shelton 1999). That the Northern Arc "would create precisely the kind of chaos that I have just escaped" was a shared sentiment amongst secessionists, who organized into a grassroots coalition to fight the road (Shipp 2002, A10). Meanwhile, to the corporate elite in Atlanta the Northern Arc would pull sprawl further outward while limiting the potential for growth in the existing urban core. The road threatened future transit projects in the urban core because it would absorb billions of dollars in federal and state funds that could be spent on transit. Again, sprawl and continued pollution from automobility threatened the exchange value and environmental health of the metropolitan region. Capitalist-financed environmental organizations, such as the Georgia Conservancy, led the fight against the Northern Arc as a proxy for the growth machine (Henderson 2004).

Lefebvre (1991) conceptualized this type of tacit coalition as a new form of spatial struggle that transcends both traditional class struggle over urban space and simplistic defence of locality, and leads towards struggles over how space is configured and utilized. Racial animosity, class difference, and moral values were set aside to achieve two diametrically opposed visions simultaneously. The secessionist automobility camp opposed the Northern Arc because it threatened their privileged spaces: their gardens and parks, their nature and greenery, and their homes. The corporate elite, augmented by environmental and social justice interests in the urban core, could set aside their previous antagonisms towards the secessionists over transit expansion, and work in unison to fight the road. In early 2004, after years of bitter contestation, this loose coalition between secessionists and urban-oriented environmental, business, and social justice organizations defeated the Northern Arc. It remains noticeably absent from Atlanta's latest transportation plans. It is doubtful this defeat would have been possible without the vocal opposition of articulators of secessionist automobility, whose spaces of automobility remain effectively unchallenged.

Spatial Implications

The transit détente reveals that automobility remains entrenched, not because it is natural or inevitable, but because the interaction between competing mobility visions have resulted in a stalemate in places like Atlanta. Corporate elites have not offered a way around secessionist automobility. Capitalist negotiation with secessionists has produced a transit détente in which, it is estimated, 80 percent of Atlanta's workers will continue to drive alone to work by 2025 but 45 percent of all travel (not just commuting) will be in congested conditions. The average Atlantan will still drive over 30 miles a day and will spend more daily time in automobiles (ARC 2003).

Congestion threatens the exchange value of the region, and Atlanta's corporate elites will no doubt seek to manage demand for roads using neoliberal, free-market strategies such as congestion pricing, parking pricing, and toll roads (see Trumper and Tomic's chapter on such strategies in Chile). The social justice implications of these policies will be challenging, with the privilege of speeding past congestion granted to those willing or able to pay, while the lower classes remain entrapped in a slower manifestation of automobility.

With no regionally impressive transit or roadway expansions, congestion will likely become a form of de facto growth management in Atlanta. As suggested above, the mobility advantages of car usage will invariably decline and as congestion intensifies, more people will seek to avoid increased daily travel times and move closer to work, shops, and urban amenities – if they can afford the move. After three decades of population decline, upwards of 10,000 people moved to central Atlanta in the late 1990s, partly out of desire to avoid congestion (Cauley 1999). A large telecommunication corporation has shuttered parts of its dispersed suburban offices and consolidated around key transit stations. A massive new mixed-use residential, retail, and office development has risen out of an abandoned steel mill adjacent to downtown Atlanta. Streetcars and extensive pedestrian infrastructure have been proposed to stitch together a limited area of the urban core of Atlanta. However, these re-centralizing developments will also have significant implications for accelerated gentrification and may be inaccessible for those who cannot afford the compact, walkable islands in Atlanta's sea of automobility. Due to the rigorous opposition of secessionist automobility, Atlanta's corporate elites will no longer propose expanding MARTA rail or building an extensive commuter rail network with compact development around stations – proposals advocated as far back as the late 1960s. Under the current transit détente, secessionists can continue to sprawl outward from Atlanta in every direction, but they will face immense congestion, trading more time driving for spatial secession.

Conclusion

This chapter has sought to show the complexity and nuance of the politics of automobility in the case of Atlanta, Georgia where diverse factions debated its future. This study suggests that a more sophisticated analysis and critique of automobility needs to replace the fatalistic assertion that a love affair makes political challenges to automobility impossible. Approaches that essentialize automobility are misleading and unconstructive; they dampen the politics of possibilities. Automobility is not just about movement or the convenience of getting from point 'A' to point 'B', nor is it adequate to conceptualize it as a neutral agent responding to consumer preference or market demand. Rather, automobility embodies deeper social conflicts.

One of these embodiments is secessionist automobility, or automobility as a medium for physical separation and physical expression of racialized, anti-urban ideologies. While some secessionists are both racist and anti-urban, not all secessionists are racist. Nevertheless they share a vision of secession from urban space, resistance to compact patterns that support transit, and abhorrence to resolving difficult urban problems through cooperation and consensus – secession by car is easier.

In seeking to mitigate air pollution and congestion, which threatened the exchange value of the region, a corporate elite in Atlanta arbitrated secessionist automobility. Articulators of secessionist automobility contested corporate elite policies of expanding transit; out of that struggle evolved a transit détente that provides a limited geography of transit service. Secessionists also stood in the way of Atlanta's highway builders, who sought to create a massive new outer beltway to spur further automobility. Ironically this positioned the secessionists, who waged what amounts to a culture war against cities, as unwitting allies of the corporate, environmental, and social justice interests who at the same time battled them over expansion of transit. Transit policy was not aimed at re-orienting everyday life for the entire region in order to reduce automobility, but rather was the result of a stalemate in a struggle, negotiated by Atlanta's capitalist growth machine to maintain the exchange value of the metropolitan region and its competitive position in the global competition between cities.

In cities throughout the world, automobility is a central site of struggle about how space is configured and for whom (Lefebvre 1991). This framework enables more clarity in efforts to address the plethora of ecological, social, and economic problems that stem from automobility. If automobility is framed in a way that focuses on what ends people are trying to achieve, rather than as an essentialized love affair, could arguments against the proliferation of automobility take a different trajectory? In the struggle over automobility, what other conceptualizations of spatial organization are deployed? How will conceptualizations centred on sustainability and social justice counter secession, or negotiate the capitalist arbitrated stalemate in cities like Atlanta? These questions call for deeper examination of how automobility is contested, for the struggle against its deleterious effects will likely intensify globally. As Martin's chapter indicates, critical mass bicycle protests and reclaim the streets movements have directly contested the right of way of automobiles, while bus riders' unions raise concerns about access to transportation, and the global middle class worries about declining quality of life. New movements that seek to reclaim urban spaces from automobiles, limit their use, and more broadly, attempt to de-emphasize high speed mobility and car ownership, will need to face the complex and nuanced political realities that arise when contesting the automobile.

References

ARC (Atlanta Regional Commission). 2003. *Transportation planning briefing book.* Atlanta, GA: Atlanta Regional Commission.

——. 2004a. *Atlanta region transportation fact book.* Atlanta, GA: Atlanta Regional Commission, p. 63.

——. 2004b. Mobility 2030: Draft for public comment. Atlanta, GA: Atlanta Regional Commission.

Atlanta Journal-Constitution. 1998. Editorial: Turning down a transit gift. *Atlanta Journal-Constitution* 30 August: A10.

Beauregard, Robert A. 1993. Representing urban decline: Postwar cities as narrative objects. *Urban Affairs Quarterly* 29(2): 187–202.

Cauley, H. M. 1999. Intown neighborhoods: Backlash against suburban sprawl spurs condo sales. *Atlanta Journal-Constitution* 16 May: H8.

Cordell, Actor. 1987. Racism called regional transit roadblock. *Atlanta Journal-Constitution* 3 July: A1.

Docherty, Iain. 2003. Policy, politics and sustainable transport: The nature of labour's dilemma. In *A new deal for transport?: The UK's struggle with the sustainable transport agenda,* Iain Docherty and Jon Shaw, eds, pp. 3–29. Malden, MA: Blackwell Publishing.

Dunn, James A. 1998. *Driving forces: The automobile, its enemies and the politics of mobility.* Washington, DC: Brookings Institution.

Freund, Peter and Martin, George. 1993. *The ecology of the automobile.* Montreal, PQ: Black Rose Books.

Garrow, David J. 2000. Rocker supporters cheer symbol of their own racial, ethnic biases. *Atlanta Journal-Constitution* 7 June: A19.

Goldfield, David R. 1982. *Cottonfields and skyscrapers: Southern city and region, 1607–1980.* Baton Rouge, LA: Louisiana State University Press.

——. 1997. *Region, race and cities: Interpreting the urban South.* Baton Rouge, LA: Louisiana State University Press.

Gordon, Peter, Richardson, Harry W. and Jun, Myung-Jun. 1991. The commuting paradox: Evidence from the top twenty. *Journal of the American Planning Association* 57(4): 416–29.

Harvey, David. 1989. *The urban experience.* Baltimore, MD: Johns Hopkins University Press.

Henderson, Jason. 2002. Contesting the spaces of the automobile: The politics of mobility and the sprawl debate in Atlanta. Ph.D. diss., Department of Geography, University of Georgia, Athens, GA.

——. 2004. The politics of mobility and business elites in Atlanta, Georgia. *Urban Geography* 25(3): 193–216.

Ihlandfeldt, Keith R. and Sjoquist, David L. 1998. The spatial mismatch hypothesis: A review of recent studies and their implications for welfare reform. *Housing Policy Debate* 9(4): 849–93.

Interview one. 2001. Personal interview with Cherokee County Commissioner, Atlanta, GA, 14 June.

Interview two. 2001. Personal interview with Gwinnett County Commissioner, Atlanta, GA, 13 June.

Keating, Larry. 2001. *Atlanta: Race, class and urban expansion.* Philadelphia, PA: Temple University Press.

Lefebvre, Henri. 1991. *The production of space.* Malden, MA: Blackwell.

Logan, John R. and Molotch, Harvey L. 1987. *Urban fortunes: The political economy of place.* Berkeley, CA: University of California Press.

Long, Bryan. 2001. GRTA awaits counties' actions. *Atlanta Journal-Constitution* 26 March: D3.

Luard, Tim. 2003. Shanghai ends reign of the bicycle. *British Broadcasting Corporation* 9 December. http://news.bbc.co.uk/go/pr/fr/-/2/hi/asia-pacific/3303655.stm (accessed 10 December 2003).

McCosh, John and Shelton, Stacy. 1999a. Teaming up with mason benchmark for Shackelford. *Atlanta Journal-Constitution* 17 September: C9.

——. 1999b. Sultan of sprawl: A profile of Wayne Hill: Part 1. *Atlanta Journal-Constitution* 19 December: C1.

——. 1999c. Profile: Wayne Hill: Part 2. *Atlanta Journal-Constitution* 20 December: B1.

McPherson, Crawford. B. 1962. *The political theory of possessive individualism: Hobbes to Locke.* Oxford: Oxford University Press.

Mitchell, Don. 2005. The SUV model of citizenship: Floating bubbles, buffer zones and the rise of the 'purely atomic' individual. *Political Geography* 24(1) January: 77–100.

Nelson, Arthur C. 2001. Exclusionary zoning, sprawl and smart growth. Spr' all come on in: Local and comparative perspectives on managing Atlanta's Growth, Presentation at Georgia State University, Atlanta, GA, 1 February.

Patton, Ken. 1998. Cherokee county land use plan: Presentation to Athens Federation of Neighborhoods, Athens, GA, 3 November.

Pearlman, Jeff. 1999. John Rocker. *Sports Illustrated* 21 December. http://sportsillustrated.cnn.com/features/cover/news/1999/12/22/rocker (accessed 15 January 2000).

Pucher, John, Peng, Zhong-Ren, Mittal, Neha, Zhu, Yi and Korattyswaroopam, Nisha. 2007. Urban transport trends and policies in China and India: Impacts of rapid economic growth. *Transport Reviews* 27(4): 379–410.

Reed, John Shelton. 2001. What's southern about the South…these days? Presentation at the University of Georgia, Athens, GA, 22 March.

Schneider, Craig. 2000. How has John Rocker changed us? *Atlanta Journal-Constitution* 12 January: D1.

Shelton, Stacy. 1998. Gwinnett businesses like bus plan. *Atlanta Journal-Constitution* 21 September: E1.

——. 1999. State still moving on Northern Arc. *Atlanta Journal-Constitution* 28 May: C2.

Shipp, Bill. 2002. Daughter disagrees that father knows best. *Athens Banner Herald* 12 May: A10.

Smith, Jay. 2000. Rockin' our world. *Creative Loafing* 28 April: 29–31.

Sperling, Daniel and Claussen, Eileen. 2004. Motorizing the developing world. *Access* 24 spring: 10–15.

Stone, Clarence. 1989. *Regime politics: Governing Atlanta 1946–1988.* Lawrence, KS: University Press of Kansas.

Texas Transportation Institute. 2004. *The 2004 urban mobility study.* College Station, TX: Texas A&M University System.

Torpy, Bill. 1999. The visionaries. *Atlanta Journal-Constitution* 26 July: E1.

Transportation Research Board (TRB). 2001. *Making transit work: Insights from Western Europe, Canada, and the United States.* Washington, DC: Transportation Research Board, National Research Council.

Urry, John. 2004. The 'system' of automobility. *Theory, Culture and Society* 21(4/5): 25–39.

Vigar, Geoff. 2002. The politics of mobility: Transport, the environment and public policy. London: Spon Press.

Vuchic, Vukan R. 1999. *Transportation for livable cities.* New Brunswick, NJ: Center for Urban Policy Research.

Ward, Janet. 2001. Investment in rail is an idea on the right track. *American City and County* April: 2.

Wood, Douglas S. 2000. For a growing Atlanta race has always mattered. *Atlanta, CNN Interactive.* http://www.cnn.com/specials (accessed 1 October 2000).

Young, Andrew. 2000. Rocker has chance for redemption. *Atlanta Journal-Constitution* 11 January: A11.

Chapter 9

The Chilean Way to Modernity: Private Roads, Fast Cars, Neoliberal Bodies

Ricardo Trumper and Patricia Tomic

Introduction

According to Sheller and Urry, the car illustrates 'putative globalization': one billion cars were produced in the twentieth century, of which 500 million still run, and these numbers are expected to double by 2015. They argue that the car dominates globally; cities are "rooted in and defined by automobility," civil society is constituted by car-drivers, car-passengers, pedestrians and "others not-in cars" (2000, 739). In sum, as "a complex amalgam of interlocking machines, social practices and ways of dwelling, not in a stationary home, but in a mobile, semi-privatized and hugely dangerous capsule ... [automobility has] ... reshaped citizenship and the public sphere via the mobilization of modern civil society" (739). Later, Urry adds that "country after country is developing an 'automobility culture' with the most significant currently being that of China," and concludes that automobility is a system in the making, emerging first in the North and taking over most of the globe (2004, 25). Indeed, he says, invoking a series of chapters in Miller (2001), "the poorer the country, the greater is the power of this virus" (27). For him, the system of automobility, auto-spreads worldwide and is self-producing. It is worthwhile reproducing his argument verbatim:

> Automobility can be conceptualized as a self-organizing autopoeitic, non-linear system that spreads world-wide, and includes cars, car-drivers, roads, petroleum supplies and many novel objects, technologies and signs. The system generates the preconditions for its own self-expansion ... automobility produces through its capacity for self-production, what is 'used by a unit as a unit.' It is through automobility's restructuring of time and space that it generates the need for ever more cars to deal with what they both presuppose and call into existence. (Urry 2004, 27)

In contrast to Urry, Gartman (2004) argues that there is a dialectical linear dynamic in the production, consumption and expansion of systems of automobility. And if class plays a significant role in this dynamic, then it is possible to position the expansion of car systems in a period of postfordist, neoliberal globalization characterized by the reconstruction of bourgeois power and the extension of market relations to every cranny of societies worldwide (Harvey 2005). This new configuration confers on global corporations an almost unlimited reach to roam the world for profits, including public goods until recently declared off limits to profit pursuits. The expansion of systems of mobility has to be understood as located in the political and economic realms, planned and connected to the interests of huge transnational combines that underwrite the system as they interact with local interests and cultural practices in places where the American dream, anchored in automobility, is still to be attained. As Martin (this volume) argues, the social and ecological ramifications of automobility take different forms around the world.

In this chapter we study the particular configuration of the Chilean system of automobility beginning from Freund and Martin's (2000) contention that complete automobility is for many the standard to emulate. Chile is not an exception. In particular, Pinochet's dictatorship, which established the first organized neoliberal experiment in the world after the 1973 coup, associated the car with development and modernity. Following Harvey (2005) we further claim that neoliberalism is a product of a class power that has expanded and deepened not only in Chile but worldwide as a global bourgeois project. The automobilization of Chile has not been the result of a system that has created the conditions for its own self-expansion. Rather, it has been propelled deliberately by the Chilean state and powerful transnationalized economic interests for capital accumulation and profit making. The key to automobilized neoliberalization is roads and urban transformation. We explore the logic of neoliberal automobility as the Chilean population and Santiago's built environment have suffered the twin impacts of the privatizatisation of pensions plans and roads. That is, we localize automobilization within the framework of a neoliberal State that has, strategically, on the one hand, allowed transnationalized capital to extract surplus from Chilean workers' pension schemes (Hormazábal Sánchez 2006), and, on the other, opened the country to giant transnational companies in the business of Build-Operate-Transfer (BOT) roads. Thus, the central actors in the Chilean automobilization scheme are the Chilean pension plans and the transnational Toll Road Companies operating in Chile.[1] Like Henderson (this volume), we argue that automobility is not inevitable, and start from the idea that the development of a burgeoning system of automobility in Chile is not the result of the natural love affair of Chileans with the car, but is rather the result of the politics of automobility.

---

1   Many of the most important highways in Europe, Asia, and Latin America are operated by no more than sixteen transnational corporations. Among these, Acciona, Abertis, ACS/Dragados, Ferrovial/Cintra all with headquarters in Spain; Autostrade from Italy; Cofiroute and Vinci from France; Hochtief from Germany, and Skanska from Sweden have had important interests in Chile.

Neoliberalism, Modernity, Transit and Automobility

The geographical proliferation of the neoliberal version of capitalism since the 1970s has been uneven. Chile was the first deliberate and organized attempt at transforming capitalism from the Keynesian form it had taken since the Great Depression into neoliberalism, preceding Thatcherism and Reaganism (Harvey 2005; Taylor 2005). Neoliberal policy and theory argue for social formations where every human relation is commodified and structured by 'free' market forces (Harvey 2005). In 1979, after four years of economic 'shock treatment' and repression, the Pinochet regime made the initial organized move in that direction. The regime granted:

> the large conglomerates, with which it meshed, power over workers and access to untapped areas to exploit. On the one hand, the regime designated the *Plan Laboral*, which curtailed workers' ability to unionize and unions' power. On the other, it moved to create the *Reforma Previsional*, a scheme that allowed private companies to seize their workers' pension contributions, around 10 percent of all employees' salaries and to appropriate another 3 percent in 'administration' fees. (Tomic and Trumper 1999, 50)

When the dictatorship launched the *Plan Laboral* and the *Reforma Previsional* it invested modernization with the authority to serve as the discursive springboard for future actions (Piñera 1991). The never-ending pursuit of '*modernización*' [modernization] became a crucial aspect of the neoliberal discourse for Pinochet and the civilian 'democratic' presidents who have followed him (Trumper and Tomic 1999).

The discursive deployment of modernization positioned the car as one of the main artifacts of modernity. In 1979, Pinochet promised that "toward 1985 or 1987, every Chilean worker will have a house, a car, and a television. Perhaps they will not have a Rolls Royce, but they will have a 1975 *Citroneta* [Citroën 2CV]" (Osorio y Cabezas 1995, 77). Seemingly, the dictatorship saw a nexus between the car and the new neoliberal individual it sought, that is, the car as a hybrid that changes drivers' views of the world, making them more consumerist, individualistic, and concerned with the pursuit of self-determined private purposes (Williams 1997).

However, the necessary infrastructure for implementing this scheme did not exist at that time. On the contrary, under Pinochet all roads deteriorated. In the 1980s, investment in infrastructure was 70 percent below the minimal requirements for the maintenance of the existing roads. This resulted in a significant decay of all roads to such an extent that only 15 percent of existing roads could be considered to be in good condition (Fuentes y Sierralta 2004; MOP 2003). To a large degree, the deterioration of the road system was a consequence of the economic crisis that hit Chile in the early part of the 1980s. The rate of poverty remained high until the end of the military regime in 1990; by then, more than 40 percent of all

Chileans lived in poverty. Car ownership grew slowly under the Pinochet regime. An influential study on transportation in Santiago shows that despite Pinochet's promises and notwithstanding the easing of restrictions on car imports, the car fleet expanded only 50 percent in the period 1977–91. In 1991 there were 400,000 cars in Santiago, or one for every 10 inhabitants. Only 19.7 percent of all trips made in the city were by car, whilst 70.5 percent were by mass transit (SECTRA 2002).

In 1990, Pinochet was replaced by a coalition of political parties (the *Concertación de Partidos por la Democracia* or simply *Concertación*). Although claiming a center-left position in the political spectrum, the *Concertación* reached an agreement with the military regime to continue with the neoliberal model of modernity as the goal for Chilean society. In fact, the *Concertación* deepened neoliberalism (Winn 2004) and redoubled the political role of the car as a symbol of modernity.

Since 1990, *Concertación* civilian governments have consolidated neoliberalism, reproducing one of the most unequal societies in the world (Kremerman 2004; Solimano y Pollack 2006). In an environment of almost continuous economic growth (except for a brief recession in 1997) (Taylor 2005), and faced with political pressures from extra parliamentary forces, the *Concertación* has been pushed to attempt to stop the existing income distribution from further deteriorating. While most sectors of the population have increased their incomes and the possibility of car ownership has extended more broadly, the automobile continues to be beyond the means of many. Pragmatically, the higher incomes are used in part to acquire automobiles. In the Chilean exceptionally classed society, the car is the cocoon used by the middle and upper classes to move within the exclusive corridors of modernity to carry out their daily lives, segregating themselves from others – a form of secessionist automobility (see Henderson's chapter in this volume). In turn, the car is also sought by the poor to flee their often harsh neighbourhoods and to reach work and leisure without having to resort to the inefficient system of mass transit. As Soron (this volume) argues, the central role played by the car in contemporary life arises "from pragmatic responses to our changing social and material environment."

From 1991 car imports have consistently increased in Chile. During the first civilian government, 1990–94, the vehicular fleet grew 32 percent. Since then, car imports have stood at more than 100,000 a year. The last large study of transportation in Metropolitan Santiago in 2001 counted around 800,000 automobiles; one car for every two families (SECTRA 2002). By the early 2000s, vehicle imports grew again, reaching almost 200,000 in 2006, when the government eliminated one of the last symbols of the car as a luxury item, a surcharge tariff on more expensive vehicles that had survived even Pinochet's regime. The same study also showed that by 2001 mass transit was losing ground as a means of transportation in the Santiago region, and that cars were playing an ever increasing role. Although the majority of trips were still on mass transit (51.9 percent), since 1991 trips in automobiles nearly doubled to 39.2 percent (SECTRA 2002).

Despite dramatic automobilization, a considerable part of the population continues to fall under the category 'others not-in cars' and to depend on mass transit. Chile has one of the most unequal income distributions in the world, in which the wealthiest 10 percent of the population makes 39.7 times more than the poorest 10 percent (Délano 2004; Kremerman 2004). In Santiago the majority of the population lives in neighbourhoods devoid of green spaces, recreational facilities, or attractive schools, crammed in narrow apartments or flimsy houses. They are relatively immobile. Public transit is expensive for their incomes and dangerous to use (Tomic and Trumper 2005). Meanwhile, wealthier families own one or more vehicles. The political division of Santiago into municipalities corresponds to not only residential segregation, but also car ownership. In the poor municipalities, cars are few and outdated. The streets are narrow; in the *poblaciones* [poor neighbourhoods] many streets are *pasajes* [alleys] and are often unpaved. In contrast, the few wealthier municipalities, traditionally to the east of the city, but now expanding north, have many parks and much greenery, the best housing, clinics and schools, the majority of corporate buildings, and broad streets to facilitate car domination. In the eastern municipalities of Santiago, where most of the rich live, the rate of car ownership is higher than one per family. For example, in the early 2000s, in Lo Barnechea, one of the wealthiest municipalities in Chile, there were 1.61 vehicles per household. In contrast, in the poorest municipalities, car density was much lower. In La Pintana, for example, in the southern part of Santiago, the rate of automobilization was around 0.3 vehicles per household (many of which are old and dilapidated, and lacking catalytic converters to combat smog, are subject to a no-drive-days policy). In Lo Barnechea, public space is built around car use; most trips are made by car (64.8 percent in 2001). In La Pintana, in contrast, in 2001 only 9.6 percent of all trips were made by private car, 36.6 percent were by mass transport, and 48.1 percent were made on foot. The class-specific use of diverse means of transportation is city wide. In 2001, mass transit was utilized by 58.1 percent of the lowest paid sectors; by 40.4 percent of middle- income people; and by only 15 percent of those with high incomes (SECTRA 2002).

The poorest strata of Chileans are mostly restricted to the municipalities where they live, except for going to work, or for the myriad of errands that most Chileans have to endure in highly bureaucratic procedures in banks, utility companies, government offices and the like. In these neighbourhoods people are often compelled to rush home before nightfall, fleeing inadequately lit public space to confine themselves to dwellings protected by iron bars and self-made cul de sacs built with elaborate improvised fences. Transportation is expensive. It takes a substantial part of people's incomes and time. Members of the popular classes who need to travel are relegated to the bus, which until recently has been the main form of mass transportation in Santiago. Until February 2007 buses were privately owned, poorly regulated, noisy and dangerous. Owned by small entrepreneurs, they operated under a system that promoted competition between different bus routes that shared the same roads. The approximately 8,000 yellow buses in Santiago created chaos on the streets (Tomic and Trumper 2005). In contrast to the buses, Santiago's publicly owned subway system was a node of modernity since its inception in the 1970s. It was clean, secure, reliable, and fast, but compared to the bus, had relatively limited reach. As people of lesser means often live far from places of work, and the mass transit system did not permit transferring between buses or to and from the subway, many people had to embark on bus trips as long as two hours each way, every day of the week. The system was drastically overhauled in February 2007; the buses and the subway were integrated, the number of buses reduced, and the small entrepreneurs replaced by a small number of large companies. However, the new system, called Transantiago, has been a dismal failure, increasing further the times of travel, nearly collapsing the subway, resulting in continuous social protests and challenging the government. In the last few years, in contrast, the most affluent have seen a major revolution in their means of transportation with the building of high-speed urban toll roads and the construction of exclusive walled neighbourhoods in the east and north of the city. Ironically, this infrastructure has been possible, in part, through the privatization of workers' pension plans.

Pensions and City Expansion

In 1979–82, under the guise of 'modernizations', the neoliberal dictatorship created markets for private capitalists in areas that had previously been out of bounds. As part of this process, the government privatized the pension system. In 1980, Decree 3500 created a new type of company, the *Asociación de Fondos de Pensiones* (AFP). Workers (but not employers) had to contribute 10 percent of each pay cheque to a retirement account in one of these companies. In addition, the AFP would charge around 3 percent of the workers' salaries for administration and disability insurance. Thus, by law, in this compulsory pension system workers have to hand over 13 percent of their income to for-profit companies in which they have no representation. In practice, after 25 years of operation of the AFP system, most workers' pensions are miserable (Rossel 2004). In the meantime, the AFPs have made enormous profits and gained power by investing workers' contributions in shares of many of the most powerful companies in the country, and in the last few years, abroad. Since its inception, ownership of the pension system has become more concentrated and transnationalized. By 2006, there were only six large AFPs left, the majority owned by transnational capital. These companies receive about US $3 billion a year in workers' deposits, and with rates of return on capital of about 30 percent a year, make immense profits that reached US $175 million in 2005 (Saleh 2006). What is germane for our purpose is that by law, a good proportion of the funds that the AFPs administer have had to be placed in Chilean fixed income investments such as bonds and mortgages. As a result, real estate investments by AFPs have stimulated a prolonged housing boom. AFP managers boast that they have lent US $5,400 million to finance the construction of 300,000 houses and hold 47 percent of all mortgages. José Piñera, the original architect of Pinochet's [seven] 'modernizations', including the AFPs, claims that two of every three houses built in Chile are constructed with AFP funds. In addition, the AFPs have invested US $500 million in so called *Bonos de Inversión Inmobiliarios* (Real Estate Investment Bonds) to finance ventures such as office buildings and malls (Asociación Gremial de Administradoras de Fondos de Pensiones N.d.; Bonet 2003). Moreover, the AFPs have a close relationship with insurance companies; many AFPs and insurance companies have common ownership. Since around two percent of the workers' contributions collected by AFPs are to be spent in purchasing disability and joint life insurance, often the AFPs buy insurance from the companies linked to them by common ownership. Thus, a significant amount of workers' income is transferred to insurance companies, boosting the conglomerates' profits and providing them with more liquidity to finance still more construction. Also, workers who receive a pension from an AFP have the choice to buy annuities by transferring their accumulated funds to insurance companies. These insurance companies, in turn, use part of these funds for *Bonos de Inversión Inmobiliarios* (Bonet 2003).

Meanwhile, in an effort to keep the existing income distribution from deteriorating further, and to feed its claim to growth with equity, among other social strategies, the *Concertación* governments have embarked on housing projects for the poor. The government has financed so-called '*vivienda social*' [social housing] to house the poor in flimsy, poorly built, overcrowded, unlighted, and dusty conditions. Complemented by poor services in education and health, and lacking amenities and parks, the '*vivienda social*' represent the Chilean version of the stereotypical urban American 'project'. In Santiago, they are placed where land is cheap, on the periphery of the city, often in formerly rural areas that have been incorporated into the urban fabric (Hidalgo et al. 2007).

Thus, the reproduction of the neoliberal model reigning in Chile for more than three decades has depended on continuous growth in real estate stock, which, in turn, propels ceaseless urban expansion beyond existing city limits. This expansion requires people to travel farther from home to work, study or play, commanding automobilized mobilities for the more affluent and mass transport for the rest.

Despite pledges by *Concertación* governments to rein in the growth of Santiago, the urban area increased from 50,000 hectares in 1990 to 62,000 in 2000, and continues to grow (Ducci 2002). Workers are semi-immobilized in neighbourhoods of social housing, in older *poblaciones*, or in crumbling houses in Santiago's historical zones, frequently close to highways or train tracks. The rich have moved into an arc of valleys or higher lands in the east and northeast of the city. The consolidation of this segregated housing order has been anchored in the contradictions of mobility and immobility, between an increasingly mobile automobilized middle and upper class, and 'others not-in cars' living close to newly built high-speed BOT roads they cannot access. Filled with cars, these roads pollute with their noise and smog. The Chilean government has developed a new system of automobility to support and ceaselessly expand this urban logic that serves as a blueprint for further motorized mobility. Shopping malls and large supermarkets, for example, are easily accessible by car. Malls in the wealthy areas are almost exclusively intended for car drivers, although buses carry those who work there, much like suburbs in Atlanta (see Henderson's chapter). Malls for the less affluent have multiple points of access, as many are close to subway stations and bus routes, in addition to roads. Santiago's expansion according to the logic of the car exemplifies Martin's (this volume) analysis of the social and ecological effects of contemporary motorization.

Private City Highways

Highway construction had a central place in the reordering of Santiago to satisfy the demands of a capital accumulation model based on forced pension contributions to private companies. The first paid highways, mostly built by transnational corporations such as SACYR and Tribasa in the mid-1990s, were intercity roads. Although partly motivated by the need to absorb increasing car and truck traffic, and to overcome the deficit left by the dictatorship's years of neglect, road construction was closely connected to the pension system, and to the transnationalization of the Chilean economy. They foreshadowed the system of private urban toll roads in Santiago, the *carreteras concesionadas* [concession roads].

The concession roads have been a decisive measure of the *Concertación* governments to strengthen neoliberal policies and ethics. Modernity, commodification and private property are linked by these roads. In this case, Chilean neoliberalism followed a procedure that was also being established in Argentina, Mexico and Europe. Western Europe today has a powerful private toll road industry that invests in Latin America, Eastern Europe, the UK, Canada and the US (Poole and Samuel 2006). Australian companies have also begun to penetrate this new market. Transnational construction engineering firms, as well as banks and pension funds, participate as partners in these ventures (Florian 2006).

Road concessions in Chile are part of the Ministry of Public Works' Concession Program for airport, road, jail, dam, and port construction, maintenance and operation (Lorenzen et al. 2003). Government and industry have deployed a number of discursive elements to explain and justify the transference of responsibility from government to private interests in these ventures, which includes transferring existing roads, bridges and airports to private and mostly transnational interests. For example, a former minister of Public Works argued, "at the time the [toll roads] program was designed, the country's most pressing infrastructure need was to upgrade and modernize the national highways, and this sector became the starting point for the concessions program" (Cruz Lorenzen et al. 2003, 1). He also stated that "private-public partnerships" would save the country's resources to invest in much needed social programs unprofitable for the private sector: "the approach, if successful, would make it possible to shift public expenditures to programs that provided high social returns but that did not offer significant investment opportunities to the private sector" (Cruz Lorenzen et al. 2003, 1). He also asserted that road privatization was needed to reduce accidents, which had doubled during the dictatorship.

What was not said by government officials is that like housing and office construction, highway building was a venue for AFPs and Insurance Companies to invest the capital that they had accumulated. As in the case of housing construction, AFPs brag about this process and prominently display billboards on highways linking contributions to private pension funds and the building of BOT roads. For example, in 2005, a billboard on a privatized route claimed "*Su Ahorro Financió Esta Carretera, Esta Carretera Financiará·Su Pensión*" ["Your Savings Funded This Highway, This Highway Will Fund Your Retirement"] (Figure 9.1). Another, in 2006, boasted "*En Esta Carretera Pusimos Tus Ahorros Para Hacer Crecer Tu Pensión*" ["We Placed Your Savings In This Highway To Make Your Pension Grow"]. However, the subsidies and legal warranties of profitability given to companies by government were hidden from public and even parliamentary scrutiny.[2]

In the 1990s, around 20 companies took over many existing roads, and using capital that, at least in part, had been extracted from Chilean workers by AFPs,

Figure 9.1   Pension Funds and Highways, 2005 – Billboard on Highway 68

widened and upgraded them at little risk but for high profits. Most of these were transnational companies in the global business of road building and toll road administration. In a minority of concessions the companies had to completely build new roads, such as the Ruta del Itata, from Chillán to Concepción in the southern part of Chile, and large segments of the Carretera del Sol, from Santiago to the coastal areas close to the port of San Antonio. The new model, based on a combination of

---

2   It is only recently that it has become clear how unclear and hidden these were (see *La Nación* 2007).

private concessions and speed, was proudly offered by the *Concertación* as proof that the (private) road to modernity initiated by Pinochet had been achieved. The privatized roads became corridors of neoliberal modernity offering the affluent high-speed geographical mobility and proximity to global modernity. When concession roads began to operate and the maximum speed had been raised to 120 km/hour, a promotional campaign distributed to road users invited them to be part of a new Chile: "*Siga por Futuro. Viaje por Chile. Con Seguridad y Confianza en las Autopistas Concesionadas*" ["Follow the Future. Travel Safely and Confidently through the Concession Highways of Chile"]. The roads were proof that Chile was finally on the road to modernity. In 2005, a television program pointed out, "in fact, we are the leaders in South America in privatizing the construction of new airports and roads," stating unequivocally: "the government lends the roads to private interests and these, in return, invest, and modernize them. In addition, the government saves millions that it can use for social projects" (Contacto TV Channel 13 2005, N.p.). The model of neoliberal automobility is thus coded as development, speed, safety, modernity and even equity.

Privatizing City Roads

The privatized inter-urban toll highways were presented by the hegemonic neoliberal project as roads to modernity. While private intercity roads were relatively common worldwide, to impose a similar project on urban spaces was more audacious. Examples around the world were fewer and the political risks were much higher because it would involve the privatization of existing city roads, huge environmental impacts, city transformations, and neighbourhood contamination and pollution. In fact, urban highway building in other cities has had brutal impacts, and social movements in such cities as Toronto and New Orleans successfully stopped highway building projects (Berman 1989; Lewis 1997). And, unlike Santiago, these were freeways.

The city of Santiago is among the first experiments worldwide in creating an urban system of paid highways. Only a few cities have attempted to establish paid urban highways. Since the early 1990s toll road companies have developed and operate nearly all of the urban expressways in Melbourne and Sydney. In France, in 1995, after a protracted battle with local residents, giant multinational Cofiroute was granted a 70-year concession to design, finance, build and operate underground tunnels in Versailles, outside Paris (Poole et al. 2005). In 2005, the city of Chicago leased the urban Chicago Skyway to a conglomerate owned by Spanish Transnational Concesiones de Infraestructura de Transportes SA (Cintra) and Australian Macquarie Infrastructure Group (Poole and Samuel 2006).

In Chile, the privatization of city roads was possible because few groups challenged this policy. There was no opposition to privatizing and 'upgrading' Américo Vespucio, the road that rings Santiago. Skanska (Sweden) and ACS (Spain) were awarded the concession of Autopista Central in September 2000, which included taking over two major existing roads. Avenida Kennedy, a broad

road in the eastern part of Santiago, was also privatized and became part of Costanera Norte, given in concession to a consortium headed by Impregilo (Italy) in April 2000. In general, a campaign that promised speed, modernity, smooth traffic flows and less pollution in return for tolls paid to private companies found little resistance.

By cutting through neighbourhoods the toll highways pollute them with fumes and noise, fragment communities, force pedestrians to walk to overpasses at distances of four or five city blocks from each other, and exclude pedestrians and bicycles (see Martin's chapter on the impact of intensified motorization on urban social ecologies). Yet, opposition to toll highways in the poorer areas most affected has been sporadic and disorganized. Instead, as Henderson (this volume) shows for Atlanta, the major organized resistance sprang up in the wealthier parts of Santiago. The first urban highway built in Chile was the Costanera Norte, inaugurated 12 April 2004, designed to increase mobility for the inhabitants of the richer municipalities of eastern Santiago. Originally planned to cut through consolidated older middle and upper-middle class neighbourhoods in municipalities close to and in the downtown, the Costanera Norte project was soon opposed by upper middle class and business interests. The solution was to sink the highway in the wealthier parts of the city, while the rest, which bordered the poorer municipalities to the west of the city, was left at street level.

Jain and Guiver argue that "the car creates different lifestyle practices around time and space, excluding many temporary or non-permanent users from participating in a variety of activities, denying 'citizenship'" (2001, 572). In contemporary neoliberal Chile, car ownership and road use can also be analyzed in terms of exclusionary citizenship rights. To keep the traffic moving at all times and to synchronize toll charges among the different companies that own Santiago's toll roads (See Table 9.1), movement and payment are coupled by a sophisticated electronic system that beams information from a single car transponder (called TAG in Chile) to a central unit, permitting a constant flux of traffic and profit to keep Chilean neoliberalism on the move. With only one TAG, the four existing concessions are able to charge their particular 'clients'. Beeping each time a car crosses underneath portals strategically placed to count every kilometre, transponders remind drivers of the bond between neoliberalism, movement, segregation, modernity, and the market. On these portals surveillance by Closed Circuit Television (CCTV) cameras help to prevent drivers who do not have a TAG on their windshields from entering the road. When the use of roads is limited to those who are able to pay, the right to mobility is compromised. We may perhaps speak of a form of 'TAG-citizenship' for Santiago. It is in this context that customers resist the privatized toll system, by not paying their bills and by entering the roads illegally, despite the massive fines that these heavily surveilled roads impose on 'TAG-non-citizens'.

Yet, exclusionary practices are contradictory with the principles of privatizing roads. Paid urban highways need to attract as many cars as possible to increase their profits. In Chile, these companies are also interested in increasing congestion.

Table 9.1   Toll City Highways

In Operation 2006	Original Concessioner	Current Concessioner 2007
Costanera Norte	Impregilo (Italy), Tecsa Constructora and Fe Grande (Chile)	Autostrade (Italy)
Autopista Central	Skanska (Sweden) and ACS (Spain)	Skanska (Sweden) and Abertis (Spain)
Vespucio Norte	Hotchief (Germany), ACS (Spain), Belfi and Brotec (Chile)	Hotchief (Germany), Abertis (Spain), Belfi and Brotec (Chile)
Vespucio Sur	Acciona (Spain) and Sacyr (Spain)	Itinere (Grupo Sacyr Spain) and Acciona (Spain)
Under Construction 2007		
Anillo El Salto Kennedy	Hotchief (Germany) and ACS (Spain)	Hotchief (Germany) and Abertis (Spain)
Radial Nor Oriente	Sacyr (Spain)	Sacyr (Spain)
Acceso Sur Santiago	Ferrovial-Agromán (Spain)	Ferrovial-Agromán (Spain)

Source: MOP 2007.

Concession roads are allowed to charge higher toll fees when there is congestion on the highways. Thus, a system touted as efficient and ecologically sound is founded on profit principles that result in congestion and contamination.

However, concession roads are essential for the Chilean system of capital accumulation, providing profits for transnational corporations and a venue for investing the retirement funds of Chilean workers. Indeed, the private toll roads complement this scheme that is also contingent on expanding the city endlessly. New urban toll highways and tunnels are being planned and built for Santiago. For example, highways and tunnels to the north of the city are opening new frontiers to giant real estate developments, complete with gated cities that can be accessed by cars that travel at high speeds, ironically, these developments offer refuge from the smog and noise of the city and highways, and from the working class and poor (Borsdorf e Hidalgo 2005; Hidalgo 2004). It is the working class and the poor who indeed pay for 'modernization'. For example, a new southern access to Santiago boasts more than 30 kilometres of highways through densely populated poor municipalities, at times at a short distance from the front doors of the dwellings (Foro Ciudadano 2006), shamelessly intruding on people's lives – a wound of noise and smog, and a reminder of the power of capital and the powerlessness of those who have no access to the toll roads, the 'others not-in cars'.

Conclusion

In conclusion, the new highways and building boom of Santiago create a need for more and more cars, and a new culture of automobility adds a new dimension to Chile's social ecology and class system. At the centre of this project are users of cars and highways: practitioners of speed, neoliberal bodies driving by themselves to the northern and eastern edges of the city, modern TAG-citizens. The highways they occupy are non-places (Augé 1995; 1996) forbidden to pedestrians, fast corridors linking nodes of modernity such as malls, glass skyscrapers, and gated communities from which the poor are barred except as servants. On the margins of neoliberal modernity stand 'others not-in cars', pot-holed streets, a historically brutal bus system, teeming sidewalks, the non-modern city of polluted air, noise, and overcrowding.

Automobilization intertwines with the model of neoliberal capitalist accumulation, a pre-requisite and a consequence of the privatization of pension plans and the facilitation of profit making by transnational corporations in the business of BOT roads. The dramatic increase in cars and the construction of roads in recent years has facilitated the expansion of Santiago: an accelerated and unequal project of housing construction that has extended the city in all directions; a process of suburbanization that makes of mobility a key element for the everyday life of business and people. A highly unequal income distribution and the relative poverty of Chileans make it impossible to foresee levels of automobility comparable to those of completely automobilized societies in Europe and North America. Until 2006, those not in cars were dependent on transit, mostly on a bus network loosely modelled on and symbolic of a market guided by perfect competition.

Santiago's bus system was characterized by chaos, noise, pollution and danger, a non-modern contrast to the automobiles and the BOT road system of corridors and nodes of modernity. Thus, at the same time that the toll highways were being built, the government of Ricardo Lagos (2000–06) began to plan the transformation of Santiago's bus transit. The new Transantiago system started to operate in February 2007. It is based on the new forms of capitalist accumulation of neoliberal Chile, that is, a few large capitalist companies receive concessions to run the bus system. Although the Transantiago is organized around and interconnected to the government-owned subway, there were no government investments or subsidies planned to make the system work. However, the failure of the Transantiago has forced the State to provide continuous support to urban public transportation following a massive political backlash against the new integrated subway (public) and bus (private) system.

Cervero has pointed out that cities which have attempted to make transit a key component of their urban fabric have in common a "calculated process of making change by investing, reinvesting, organizing, reorganizing, inventing, and reinventing" (1998, 3) (and see Dennis and Urry's chapter for a discussion of innovative designs for bus services in such cities as Curitiba, Brazil). The Transantiago has none of those components. Like all the Chilean services for the powerless, this was thought of as a system on the cheap. As in education and health, where the poor are offered substandard services and the middle and upper classes receive first class assistance, the basic idea behind the creation of the Transantiago is to consolidate a dual transportation service: an automobilized one for those who can afford cars and travel through the concession roads, and a substandard private and for-profit mass transit for others not-in cars. An under-funded Transantiago comes to reaffirm and consolidate Chile's system of mobility as a class project.

References

Asociación Gremial de Administradoras de Fondos de Pensiones. N.d. Mito c.-: Los fondos de pensiones no contribuyen al desarrollo de Chile. http://www. afp-ag.cl/publicaciones/mito_c.pdf (accessed 20 May 2007).

Augé. Marc. 1995. *Non-places: Introduction to an anthropology of supermodernity.* London: Verso.

——. 1996. About non-places. *Architectural Design* 66(121): 82–83.

Berman, Marshall. 1989. *All that is solid melts into air: The experience of modernity.* New York: Penguin Books.

Bonet, Carlos. 2003. Emisión de bonos de infraestructura en Chile: Una experiencia exitosa. Feller Rate Visión de Riesgo, Junio. http://www.feller-rate.cl/general2/ articulos/ infraestructuravr0306.pdf (accessed 12 February 2007).

Borsdorf, Axel e Hidalgo, Rodrigo. 2005. Los mega-diseños residenciales vallados en las periferias de las metrópolis latinoamericanas y el advenimiento de un nuevo concepto de ciudad: Alcances en base al caso de Santiago de Chile. *Scripta Nova. Revista Electrónica de Geografía y Ciencias Sociales* IX(194)03. http://www.ub.es/geocrit/sn/sn-194-03.htm (accessed 1 August 2007).

Cervero, Robert. 1998. *The transit metropolis: A global enquiry.* Washington, DC and Covelo, California: Island Press.

Contacto TV Canal 13. 2005. Concesiones en carretera. http://contacto.canal13. cl /contacto/html/Reportajes/Concesiones/Iprofileqdenuncias.html (accessed 4 January 2007).

Cruz Lorenzen, Carlos, Barrientos, María Elena with Babbar, Suman. 2003. Toll road concessions: The Chilean experience. *PFG Discussion Paper Series*, No. 124. http://siteresources.worldbank.org/INTGUARANTEES/Resources/ TollRoads_Concessions.pdf (accessed 21 December 2006).

Délano, Manuel. 2004. Santiago de los extremos. In *Zapping al Chile actual: Serie Nosotros Los Chilenos, Vol. 1,* Manuel Délano, Tomás Moulian, Darío Oses, and Richard Vera, eds, pp. 6–27, Santiago: LOM.

Ducci, María E. 2002. Area urbana de Santiago 1991–2000: Expansión de la industria y la vivienda. *EURE Revista Latinoamericana de Estudios Urbano Regionales* 28(85): 187–207.

Florian, Mark. 2006. Public-private partnerships: An alternative source of capital municipal and infrastructure finance group. Goldman, Sachs and Co.

19   October.   http://www.surfacecommission.gov/Mark%20Florian_9.ppt. (accessed 14 January 2007).

Foro Ciudadano. 2006. Acceso sur a Santiago: El lunar más canceroso del sistema. *Diario Electrónico. Radio Universidad de Chile.* http://www.radio.uchile.cl/ notas.aspx?idNota =29034 (accessed 20 May 2007).

Freund, Peter and Martin, George. 2000. Driving south: The globalization of auto consumption and its social organization of space. *CNS* 11(4): 51–71.

Fuentes, Luis y Sierralta, Carlos. 2004. Santiago de Chile: ¿Ejemplo de una reestructuración capitalista global? *EURE. Revista Latinoamericana de Estudios Urbano Regionales* 30(91): 7–28.

Gartman, David. 2004. Three ages of the automobile: The cultural logics of the car. *Theory, Culture and Society* 21(4/5): 169–95.

Harvey, David. 2005. *A brief history of neoliberalism.* Oxford: Oxford University Press.

Hidalgo, Rodrigo. 2004. De los pequeños condominios a la ciudad vallada: Las urbanizaciones cerradas y la nueva geografía social en Santiago de Chile (1990–2000). *EURE. Revista Latinoamericana de Estudios Urbano Regionales* 30(91): 29–52.

Hidalgo, Rodrigo, Zunino, Hugo y Alvarez, Lily. 2007. El emplazamiento periférico de la vivienda social en el área metropolitana de Santiago de Chile: Consecuencias socio espaciales y sugerencias para modificar los criterios actuales de localización. Paper presented at IX Coloquio Internacional de Geocritica, Porto Alegre, 28 de mayo –1 de junio, Universidade Federal do Rio Grande do Sul. http://www.ub.es/geocrit/9porto/hidalgo.htm (accessed 28 May 2007).

Hormazábal Sánchez, Ricardo. 2006. El sistema de AFP: Un legado inaceptable. *Central de Documentos de Análisis sobre la Reforma Previsional. INAP Universidad de Chile.* http://www.inap.uchile.cl/instituto/cdrf/hormazabal04. html (accessed 15 January 2006).

Jain, Juliet and Guiver, Jo. 2001. Turning the car inside out: Transport, equity and environment. *Social Policy and Administration* 35: 569–86.

Kremerman, Marco. 2004. Distribución del ingreso en Chile: Una bomba de tiempo. *Análisis de Políticas Públicas* 29, Agosto.

La Nación. 2007. Naranjo versus Bitrán: Las dos caras de la batalla por el TAG. *La Nación*, Miércoles 3 de Enero. http://www.lanacion.cl (accessed 4 January 2007).

Lewis, Tom. 1997. *Divided highways: Building the interstate highways, transforming American life.* London: Viking Penquin Books.

Miller, Daniel, ed. 2001. *Car cultures.* Oxford and New York: Berg.

MOP (Ministerio de Obras Públicas). 2003. Sistema de concesiones en Chile 1990–2003. http://www.moptt.cl/documentos/Documento%20Concesiones% 20Final.pdf (accessed 4 January 2007).

———. 2007. Coordinación General de Concesiones. Gobierno de Chile.

Osorio, Victor y Cabezas, Iván. 1995. *Los hijos de Pinochet.* Santiago: Planeta.

Piñera, José. 1991. *El cascabel al gato: La batalla por la reforma previsional.* Santiago: Zig Zag.

Poole, Robert W. and Samuel, Peter. 2006. The return of private toll roads. *Public Roads* 69(5): N.p. http://www.tfhrc.gov/pubrds/06mar/06.htm. (accessed 14 January 2006).

Poole, Robert W., Samuel, Peter and Chase, Brian F. 2005. Building for the future: Easing California's transportation crisis with tolls and public-private partnerships. Reason Foundation, January. http://www.reason.org/ps324.pdf (accessed 22 May 2007).

Rossel, Eduardo. 2004. Viejos y miserables, *La Nación*, 8 de agosto. http://www.lanacion.cl (accessed 2 June 2008).

Saleh, Robert. 2006. La guerra de las AFP: Episodio I, *La Nación*, 9 de abril. http://www.lanacion.cl (accessed 15 January 2007).

SECTRA (Secretaria Interministerial de Planificación de Transporte). 2002. *Encuesta origen destino de viajes 2001 (EOD 2001)*. Santiago: Gobierno de Chile.

Sheller, Mimi and Urry, John. 2000. The city and the car. *International Journal of Urban and Regional Research* 24(4): 737–57.

Solimano, Andrés y Pollack, Molly 2006. *La mesa coja: Prosperidad y desigualdad en el Chile democrático.* Santiago: Colección CIGLOB.

Taylor, Marcus. 2005. Globalization and the internationalization of neoliberalism: The genesis and trajectory of societal restructuring in Chile. In *Internalizing globalization: The rise of neoliberalism and the decline of national varieties of capitalism,* Susanne Soederberg, Georg Menz and Philip G. Cerny, eds, pp. 183–99. New York: Palgrave Macmillan.

Tomic, Patricia and Trumper, Ricardo. 2005. Powerful drivers and meek passengers: On the buses in Santiago. *Race & Class* 47(1): 49–63.

———. 1999. Neoliberalism, sport and the Chilean Jaguar. *Race & Class* 40(4): 45–63.

Urry, John. 2004. The 'system' of automobility. *Theory, Culture and Society* 21(4/5): 25–39.

Williams, Raymond. 1997. Mobile privatization. In *Doing cultural studies: The story of the Sony Walkman,* Stuart Hall, Hugh MacKay, Keith Negus, Linda Janes and Paul Du Gay, eds, pp. 128–29. London: Sage Publications.

Winn, Peter, ed. 2004. *Victims of the Chilean miracle: Workers and neoliberalism in the Pinochet era, 1973–2002,* Durham, NC: Duke University Press.

Chapter 10

Driven to Drive: Cars and the Problem of 'Compulsory Consumption'

Dennis Soron

As an iconic commodity of the consumer age, the private automobile has become intimately associated with prevailing ideals of personal mobility and individual freedom. Ironically, the increasing 'automobilization' of contemporary life has also transformed this enchanted machine into a unique window onto a number of nagging problems lurking under the shiny surfaces of capitalist society – including social atomization, time poverty, stress, tragic injuries and fatalies, deteriorating public health, ecological destruction, social exclusion and the evisceration of public space. Currently, the car sits precariously in our imagination at the juncture of freedom and compulsion, serving as both a badge of individual empowerment and as a condensed symbol of many troubling aspects of our collective life that we seem impotent to change.

This chapter seeks to complement the work already undertaken on the social and environmental implications of automobility by bringing it into a critical engagement with the notion of 'compulsory consumption', as set forth in the work of Lodziak (2000; 2002), Sanne (2002; 2005) and other thinkers. Resisting the tendency within many academic and popular critiques of consumerism to regard consumption practices within 'affluent' societies as primarily culturally driven phenomena, such authors argue that the material landscapes, social infrastructures and routine pressures underpinning everyday life in such societies are the most significant drivers of unsustainable mass consumption today. To borrow the terminology used by Conley in his chapter, their analyses of consumer behaviour place much less emphasis upon the 'magical' realm of advertising, consumer ideology, social emulation and the subtle mechanisms of desire and identity construction than upon the practical and often 'mundane' pressures woven into the social environments and power structures in which consumers are embedded.

Automobile-centred transportation, I argue, presents us with an excellent example of the 'compulsory' quality of much private consumption in the advanced capitalist world today. Accordingly, challenging the hegemony of this coercive system will require much more than efforts to transform the everyday behaviours, beliefs, identities, desires and emotions of individual consumers. Deconstructing auto advertisements, interrogating the dense and varied symbolic meanings we invest into cars, making personal efforts to drive less, developing a less auto-dependent lifestyle, and so on, all usefully cast into critical relief the forms of psycho-cultural 'compulsion' underlying people's attachment to cars. In everyday social environments heavily biased towards automobile use, however, even the most conscientized opponent of car culture will continue to experience a different, externally arising 'compulsion' to drive. Of course, given the substantial variation in transportation systems between contemporary societies – and even between cities within particular national or regional settings – this compulsion can be felt to a greater or lesser degree. Commuters in relatively transit, pedestrian and bicycle friendly cities such as Amsterdam, Copenhagen or Portland, for instance, can readily avail themselves of transportation options that are much less accessible for their counterparts in heavily car-dependent cities such as Atlanta, Houston or Auckland. By more thoroughly addressing the 'mundane' material and institutional factors that exacerbate the locked-in quality of our relationship to the car, we can begin to develop a better understanding of the reasons why relatively affluent consumers in auto-dependent regions have thus far resisted undertaking any significant behavioural changes as the damaging consequences of their lifestyles have become increasingly acute. Moving beyond the very privatism that the car itself has helped to entrench at the heart of contemporary life, we can also begin to develop a more explicitly collective, political response to the problem of automobile dependency and its distinctive commodification of legitimate human needs for mobility, sociability and access.

Dangerous Driving: Cars and Over-Consumption

The historical ascendancy of the private automobile is closely entwined with the rise of consumer culture and with the general problem of over-consumption that characterizes the advanced capitalist world. "Automobility helped usher in consumerism," writes Alvord (2000, 21), "allowing the car and consumer culture to roll hand-in-hand through the rest of the twentieth century." Unfortunately, this giddy road-trip increasingly seems to be leading us towards a brick wall. As the United Nations Development Programme's 1998 Human Development Report famously attested, "runaway growth in consumption" (UNDP 1998, 2) over the past 50 to 60 years has raised material living standards for a sizable minority of the international community, but has done so at the cost of placing severe strain upon global ecosystems, sharpening inequality and social exclusion, and posing significant new problems for community life and public health. As the report outlines (UNDP 1998), combined public and private consumption expenditures worldwide grew more than six-fold between 1950 and 1998. Historically, this huge growth has been reflected in the skyrocketing consumption of basic materials such as fossil fuels, metals, foodstuffs, water, wood, paper, chemicals, and plastics, and in the relentless generation of new categories of consumer goods – telephones, televisions, home appliances of all kinds, automobiles, personal electronics, self-care products, and so on – that have gradually acquired the status of social necessities among more privileged segments of the global population.

While aggregate statistics may tempt us to generalize blame for over-consumption onto humanity as a whole, global consumption increases have consistently outpaced overall population growth and have been markedly concentrated within the wealthy industrial world. Far from reaping the bounty of the global consumption boom, a team of Worldwatch Institute researchers have estimated, 40 percent of the world's population at the outset of the new millennium struggle to survive on less than two dollars per day (Gardner et al. 2004). Conversely, as the UNDP (1998) estimates, the richest 20 percent of the global population living in a handful of high-income countries accounts for roughly 86 percent of global consumption expenditures, including an overwhelming share of the world's energy use and automobile fleet. As such figures suggest, confronting our emergent global environmental crisis in a just and equitable manner will require a substantial transformation of dominant consumption patterns within wealthy countries, and the development of less materially-intensive and wasteful methods of meeting a whole variety of collective needs, within transportation and many other spheres of life (see Litman's chapter for a discussion of transport policy alternatives).

The dizzying growth in rates of automobile ownership and usage among the 'global consumer class' (Durning 1992) over the past several decades is an important index of the ways in which capitalist social and economic development has transformed human needs and the means available to satisfy them. A century ago, car ownership was largely a luxury reserved for the periodic outings of a relatively small elite; today, in contrast, it is an everyday necessity for the great majority of the population in wealthy countries. By virtually every measure – per capita rates of auto ownership and mileage driven, the number of vehicle trips, both in absolute terms and as a proportion of total trips, and so on – the automobile has increasingly entrenched itself as the dominant mode of transportation in more developed countries (MDCs) and is beginning to do so in less developed countries (LDCs) (see Martin's chapter). In 1950, there were around 70 million motorized vehicles of all kinds in the world (Simms 2005). Just over five decades later, the Worldwatch Institute's Michael Renner calculates, the world's roads contained approximately 603 million passenger cars and 223 million commercial vehicles – with the United States, Canada, Western Europe, and Japan alone accounting for over two-thirds of this massive global fleet (Renner 2006). During the same time frame, Renner (2005) estimates, the total distance travelled by all vehicles in the United States grew more than sevenfold to roughly 4,281 billion kilometres, or the equivalent of 14,308 round trips between the earth and the sun.

As the roads and public spaces of wealthy regions become saturated with more cars making more trips, and automobile-oriented transportation gains a stronger foothold in newly industrializing countries (see Trumper and Tomic's chapter on Chile), such historical trends will only intensify. Working from the projections of the US Department of Energy, environmental analyst Ron Nielsen (2006) predicts that the global fleet of passenger cars will pass one billion by the year 2014. If motor vehicle density throughout the world were equivalent to that found in industrialized countries today, he argues, this fleet would comprise an astounding four billion cars. Such alarming predictions foretell the potentially devastating consequences of extending the current transportation norms of the global North across the entire planet.

From a consumption angle, the relevant issue here is not simply the unsustainable rate at which certain populations purchase, consume, and replace or discard cars – although built-in obsolescence, stylistic, and technical innovations, and flexible financing and leasing schemes have definitely compounded this problem. As Urry (2004) has powerfully underlined, the automobile is not simply a discrete item of individual consumption, but the linchpin of a whole system of interlinkages between central sites of production and consumption in our world. By any standard, automobiles are absolutely integral to contemporary capitalist society, facilitating economic growth and spurring continual expansion in a variety of economic spheres, from retailing and fast food to energy, road maintenance, construction, mining, and industrial, commercial, and residential land development. Cars are not only the second-largest consumer expenditure after housing for the majority of people in the West today, but a key driver of aggregate consumption of fossil fuels, aluminum, steel, iron, lead, platinum, rubber, plastics, electronics and many other goods (Freund and Martin 1993; Urry 2004). Huge amounts of collective resources are devoted to maintaining the physical infrastructure and social support systems required for pervasive, uninterrupted automobile traffic. The centrality of the car to advanced capitalist countries means that the employment, and hence consuming power, of large numbers of people is tied in with the fate of the auto industry and the many public and private economic sectors with which it is linked. Finally, of course, the unprecedented mobility that motorized transportation offers serves as a kind of circulatory system for contemporary capitalism, distributing fresh waves of goods to store shelves across vast geographical spaces, and providing the major means by which people shuttle between work, home, and various nodes of consumption – restaurants, malls, gas stations, movie theatres, big-box outlets, and so on.

As Wachtel (1983, 32) has written of the automobile, "few products have had as powerful a role in shaping the way we live, for both good and ill, and few so strongly define and limit what options appear available to us." The conundrum that relatively affluent citizens in car-dependent societies now face is that they cannot maintain their sense of entitlement to the consumerist lifestyle the car enables without also locking themselves into the troubles it leaves in its wake. In this era of smog alerts, energy and climate instability, rage-inspiring congestion, and soul-deadening sprawl, such troubles are increasingly hard to ignore. Car-oriented transportation systems are extraordinarily inefficient, meeting collective mobility needs in individualized ways that involve tremendous overcapacities, and requiring disproportionately high rates of material input, pollution, waste, dedicated space, and disruptions to nature, wildlife, human security and community health. Most obviously, automobile travel is an inherently energy-intensive form of transportation, typically requiring up to three times the per capita fuel use of public transit (Sawin 2004). The prevalence of this mode of travel helps us to explain the fact that, by 2002, the US automobile fleet alone was consuming over 8.3 million barrels per day of oil and generating more carbon emissions than the entire Japanese economy (Renner 2005).

Aside from its thirst for fossil fuels, and its correspondingly large contribution to greenhouse gas emissions and global warming, the automobile is also, directly and indirectly, a major source of toxic air and water pollution, and a significant cause of health problems as diverse as cancer, heart disease, obesity, hypertension, asthma, lung disease, and other respiratory illnesses. Over the past century, Simms (2005) writes, car accidents have claimed over 30 million human lives worldwide, escalating steeply over this period to approximately 1.2 million global fatalities per year at present, and threatening to become one of the world's top three causes of death and disability within the next 15 years (see chapters by Wetmore and MacGregor on social and technical responses to death and injury in motor vehicle crashes). As grim as these numbers are, they tell only part of the story, failing to properly account for the even greater death-toll each year attributed to automobile-related pollution and toxic releases (Sawin 2004). Accommodating our lives to the demands of mass automobility and its requisite sprawl has also gone hand-in-hand with the reckless destruction of local agricultural land, wilderness, and wildlife, and the erosion of human-scaled, walkable communities with vibrant street life, easily accessible independent shops and services, parks and other convivial public spaces. In light of all of the problems to which our current transportation practices are cumulatively giving rise, argue Michael Brower and Warren Leon of the Union of Concerned Scientists (1999, 86), people of conscience increasingly find it "hard to avoid the conclusion that personal use of cars and light trucks is the single most damaging consumer behaviour."

Automotive Attachment Disorders

As Brower and Leon acknowledge, such flickerings of guilt and discomfort have thus far not led to any noticeable change in consumer behaviour. In spite of the manifest flaws of current transportation patterns, private cars and trucks continue to move more people across greater distances than ever before. As Simms (2005, 126) writes of this paradoxical situation:

> Cars cover and suffocate our lives like black fly [sic] on nasturtiums. Yet, spellbound, we embrace the great destroyer and design our lives, communities, and countryside around them. We welcome them into our lives when, rationally, we should be emblazoning them with public health warnings in the same style as cigarette packets.

For critics such as Simms, the yawning gulf between the sheer scale of the social and environmental ills fostered by automobile dependency and the feeble public response to them is indicative of the unshakeable, irrational attachment that consumers in wealthy regions now have to cars and car culture more generally. This idea is reflected in the by-now clichéd claim that contemporary consumers are in the throes of a 'love affair' with their cars – an affair, it seems, whose fervour no amount of abuse can dampen. "Since we love our cars with such passion," writes Zuckerman, "it is only natural that we shut our eyes to their flaws" (quoted in Alvord 2000, 7). In a similar but more theoretically sophisticated vein, Sheller (2004) argues that re-evaluating the ethical dimensions of our transportation choices can only take place after we have fully contended with the highly complex and powerful 'automotive emotions' that now bind us tightly to the car.

Quite often, commentators attribute this psycho-cultural attachment to automobiles and other consumer items largely to the influence of advertising. The great success of the multi-billion dollar advertising industry, Simms believes (2005, 129), has been its ability to make cars "psychologically indispensable" to us, investing these objects with glamour, magic, identity value, social meaning and intense libidinal appeal. In their subtle analyses, Conley (this volume) and McLean (this volume) each focus on the cultural codes and themes that automobile advertising evokes, demonstrating how such codes help both to enhance the mass appeal of private vehicles and to symbolically reconcile many of the contradictions that the system of automobility generates. The symbolic and emotive dimensions of automobility are undeniably culturally significant, mediating and shaping our relationship with cars, nature, public space, and the contested, divided, and unequal social world of which we are part. In the process, as Litman emphasizes in his chapter in this collection, cars have insinuated their way into our culture's processes of status distinction and social mapping, becoming important public markers of status and prestige.

Since car advertisements are everywhere in contemporary culture, dominating the electronic media, sprouting up on billboards and washroom walls, weaving their way onto movie screens, and filling the pages of mass distributed newspapers and magazines, the symbolic value of cars is incontestable. Taken together, these ubiquitous advertisements constitute not only a giant marketing tool, but a loud, relentless storyteller about the car and its place in our lives. This massive discourse, however, ultimately provides us with a decidedly one-sided view of automobiles, obfuscating their cumulative impact upon our lives. Circumventing any critical reflection on the dark side of automobility, advertising speaks in nuanced and subtle ways to our latent individual hopes and desires, presenting us with idyllic images of the natural landscapes and social relationships the car itself has helped to disfigure and destroy (as both Conley's and McLean's chapters point out). To this extent, as Simms (2005, 130) somewhat grandly claims, automobile advertisements are our culture's "equivalent to Stalin's state propaganda posters, news reels, and artwork of happy smiling workers. They hide a brutal reality and a naked, dissolute Emperor."

For some thinkers, our attachment to automobiles has become so delusional and self-defeating that it must be described in terms of addiction, psychopathology or even moral depravity. As journalist Mark Hertsgaard has written, the car "is nothing less than addictive for human beings. Like cigarettes, cars are a source of seductive pleasure that eventually comes to enslave its users" (quoted in Alvord 2000, 235). Unfortunately, our car addiction has proven to be particularly hard to kick, and the heartening shift in societal attitudes towards smoking has not been accompanied by comparable changes in attitudes toward driving. Even authors with an otherwise strong critique of the systemic roots of automobile dependence, such as Newman and Kenworthy (1999), occasionally lapse into more individualized explanations that equate current patterns of car-use to forms of pathological consumption such as eating disorders and drug abuse. In all of these three cases, they suggest (1999, 60), over-consumption takes root as an inner compulsion when "we no longer exercise any conscious discretion and become addicted or develop a physical dependence problem." For Wachtel (1983, 32), our relationship to the car is the prime example of how deeply rooted "psychological mechanisms of denial" in consumer society prevent us from acknowledging the stark reality in front of us and taking stock of the consequences of our own actions. In his view, we are so irrationally wedded to the illusion that our cars are magical talismans of freedom, independence and libidinal fulfillment that we simply dissociate from the disenchanted daily experience of gridlock, smog and carnage on the road (1983, 36–7).

Taking on a moralistic rather than clinical tone, Kunstler argues in his best-selling book *Home from Nowhere* (1998) that bald complacency sits at the heart of our resistance to changing our transportation habits. Ignorance, he asserts (1998, 65), is no longer a legitimate excuse for our "wicked" behaviour; indeed, "[w]e have all the information we need to persuade us by means of rational argument that using cars the way we do is catastrophic. The trouble is we don't care." Ultimately, he writes (1998, 66), the only conclusion to be drawn is that we simply don't want to think about the damage cars are causing: "We lack the will to reflect, and perhaps the requisite virtue to acquire the will. We're too comfortable munching Cheez Doodles on the freeway right now to think about the consequences of continuing this behaviour." A more benign version of the same moral argument has recently been offered by Princen in *The Logic of Sufficiency* (2005). We have come to feel so entitled to the speed and convenience of car use, he argues (2005, 294), that even the most socially and environmentally conscientious among us "operate for the most part as if there is not a problem of automobility. Intellectually, everyone is well aware of the automobile's impacts but in practice, its use, including its ever-increasing use, is taken as a given." The fact that so few people are willing to experiment with reduced auto use, in this sense, symbolizes how habituated we have become to consumer excess, and how hard we will need to work to put alternative ethical principles of sufficiency and self-restraint into collective practice.

Automobile Dependency and 'Compulsory Consumption'

While anti-car activists and environmentalists are rightly critical of the automobile's baneful effects on contemporary life, such problems do not simply derive from the consumer's predilection for the ease and convenience of driving, and his or her emotional attachment to cars. Instead, my premise here is that the underlying problem of automobile dependency is best understood as a practical imperative in locations where the car has edged out alternative forms of transportation as viable options for meeting people's varied and interconnected transportation and survival needs. In this sense, our reliance on the automobile illustrates the many ways in which our everyday behaviours do not transparently express our consumer preferences and values – whether 'sovereign' or culturally induced. Instead, such behaviours reflect a constellation of pragmatic incentives and constraints that often propel us into making environmentally and socially problematic choices, regardless of how we feel or what we desire. Even those of us with severe misgivings towards auto-centric transportation may find it quite difficult to opt out of it, to the extent that we are immersed in social and material circumstances that presuppose, encourage and often enforce private car use. Frequently, for example, people residing in highly auto-dependent regions are unable to find employment or even shop for food unless they have access to a car.

In this sense, automobile dependency offers an excellent example of the 'compulsory' or 'locked-in' quality of much contemporary consumption, as addressed elsewhere by thinkers such as Lodziak (2000; 2002) and Sanne (2002; 2005). In the eyes of Lodziak and Sanne, many of our key consumption habits today – including driving – stem from pragmatic responses to our changing social and material environment, rather than from cultural factors such as advertising, ideological manipulation, the proliferation of extravagant new wants and desires, the erosion of non-consumerist values, and the heightened role that consumer goods have come to play in the process of identity construction. Such factors may have an influence on our choices, but I would argue that their effect on our overall volume of consumption and its impacts has often been seriously overstated by thinkers in the fields of sociology and cultural studies, and by environmental and anti-consumerism activists. In his critique of culturally oriented explanations of consumer behaviour, Lodziak (2002, 5–6) writes that the interactions between individuals and consumer culture "may help to explain an individual's preference for a Peugeot, for example, over other brands of cars, but it does not explain why an individual buys a car in the first place, or prefers a car to a bicycle as a primary means of private transport."

Addressing such questions requires us to challenge the notion that consumption is a realm of freedom in which people simply follow their own voluntaristic preferences and values (see Litman, this volume, who emphasizes the importance of prestige in individual car consumption). In this regard, the problem with neoclassical economics and mainstream green thinking alike, Sanne (2002, 273) asserts, is that their accounts of consumption,

> remain within a tradition focused on individual choice. They obscure how consumers' choices are affected by structural factors in society such as working life conditions, urban structure, and everyday life patterns. The focus on the

consumer also fails to pay attention to how producers and businesses construct the field of consumption to satisfy their interests. It neglects the role of the state and how business tends to co-operate with or pressure governments to create conducive conditions for increasing consumption.

The underlying social structures, power relations and material environments in which our everyday lives are embedded shape and constrain our choices in a variety of ways, ensuring that we do not always have immediate control over what we consume or the damage that it causes. Structural features of the social environments in which we attempt to survive and reproduce ourselves can impose patterns of consumption upon us that we have not freely chosen, fostering practical dependence upon particular commodities, and providing us with few resources and opportunities for acting otherwise. If rates of material consumption have continued to rise far past the limits of sustainability, this should not be attributed primarily to the insatiable appetites of ordinary consumers. A more adequate explanation would place greater emphasis upon institutional imperatives underwritten by political and economic power, which cause legitimate human needs to be met in an unnecessarily wasteful and destructive – but economically profitable – manner.

At the most basic level, contemporary physical landscapes provide a good illustration of Lodziak's claim (2002, 93) that we are not so much ideologically enticed as "materially manipulated" into compliance with prevailing consumption patterns. Indeed, as Freund and Martin argue (1993, 11), "the very social organization of space that auto-centered transport fosters helps to further auto dependence and to mask any sense of realistic alternatives to automobility." Over the past several decades, the refashioning of our communities and natural landscapes to accommodate growing volumes of automobile traffic has increasingly made it rational, from a purely individual point of view, to use the car as one's primary or even exclusive mode of transportation. Thus, while automobiles are, as Litman's chapter reminds us, 'positional goods' whose value is derived to some extent by their symbolic status relative to available substitutes, their cultural prestige in many cases corresponds with the real and undeniable practical advantages that they confer in certain social environments. Indeed, while remaining car-free in a dense, mixed-use European city with a varied transportation system is neither highly stigmatizing nor personally incapacitating, those without access to a car in a typical sprawling, auto-dependent North American city are likely to suffer from a loss in status that is rooted in radically curtailed access to a wide variety of social goods, amenities and opportunities.

In this light, the enhanced speed and mobility offered by cars has simultaneously generated new forms of material constraint and social exclusion, providing a spur to the spatial dispersal of human settlements and the segmentation of key social activity sites (this process of spatial dispersal, as Trumper and Tomic's chapter illustrates, can become central to state modernization projects). To this extent, as Urry (2004, 28) has argued, the very "flexibility" that automobility offers is a highly coercive one, congealing into a peculiar material environment that requires car-use for even the most basic everyday activities:

> Automobility divides workplaces from homes, producing lengthy commutes into and across the city. It splits homes and business districts, undermining local retail outlets to which one might have walked or cycled, eroding town-centers, non-car pathways, and public spaces. It separates homes and leisure sites often only available by motorized transport. Members of families are split up since they live in distant places involving complex travel to meet up even intermittently.

By its very nature, automobile transportation demands exponentially larger amounts of physical space than other forms of movement, sequestering off huge amounts of land for roads, highways, intersections, ramps, medians, barriers, bridges, on-street parking and parking lots, driveways, garages, service and inspection stations, and other automotive amenities. This voracious demand for space explains why automobile-related infrastructure and services now eat up roughly half of all available urban land area in some Canadian cities (Statistics Canada 2006). Beyond the city proper, car-oriented development, and restrictive single-use zoning patterns have given rise to disjointed landscapes of sprawling, low-density suburbs, and peripheral business areas, intersected by multi-lane expressways, strip-malls and big-box outlets fringed with massive parking lots (for the role of concrete, see Simons, this volume). In such environments, walking and cycling is often wildly impractical, time-consuming and unsafe, if not impossible, and public transit is generally inadequate and expensive to maintain. In this light, as a report published in 1997 by the US Department of Transportation put it, our "love affair" with cars "may actually be a marriage of convenience. Contemporary land use patterns require the use of private vehicles, whether or not we love those vehicles" (quoted in Sheehan, 2001, 18).

Contrary to our culture's frequent association of private car-use with self-sufficiency, the viability of automobile transportation is heavily reliant upon state intervention and the generous provision of collective resources. Indeed, this 'independent' personal activity is entirely dependent upon political decisions concerning land development, zoning, urban planning, transportation policy and investment that make driving cheap and convenient and systematically marginalize alternative forms of transportation. In his analysis of the politics of automobility, Henderson (this volume) shows, for example, how political decisions to expand motorways to help create and sustain new suburbs were intertwined with the desire of households to secede from the 'problems' of urban spaces and to prevent the development of public transportation. Such political decisions result in countless uses of public resources to facilitate automobile dependency: the construction and maintenance of roads and highways; subsidies and incentives for the auto, petroleum, construction and housing industries; auto-related health care and environmental cleanup costs; funds dedicated to licensing, traffic regulation, parking provision and enforcement, and other services; not to mention the indirect costs involved in building and protecting cross-continental pipelines and super-tankers, and maintaining armed troops in foreign lands. According to some estimates, the direct costs of automobile transport not covered by drivers themselves come in at 5 percent or more of the Gross Domestic Product of advanced industrial countries today (Sawin 2004).

Such massive public supports have helped to create a transportation system in which most of the practical benefits and incentives – in terms of time, cost, convenience, safety, comfort, flexibility, and so on – accrue to drivers at the expense of others. Because society as a whole absorbs such a large portion of the costs associated with driving and has kept the variable costs of operating a vehicle artificially low, many people's reliance on the car has become the most rational way of meeting their individual transportation needs. This is particularly true for large households and those with complex travelling patterns, since the cost and duration of journeys involving additional passengers or multiple destinations is much lower for cars than for public transit or intercity bus or train service. As many critics have noted, the main measure used for evaluating the effectiveness of transportation policy in previous decades has been the ease and convenience of drivers, with the needs of non-drivers scarcely factoring into the equation (Handy 2006; Litman 2001). Accordingly, public agencies have generally dealt with the mounting problems associated with the automobile through measures designed to prop up and expand our already overstretched auto infrastructure. The flip side of this institutionalized bias towards auto-centred transport is the lack of collective resources and policies now devoted to making alternative modes of transportation more convenient, safe, affordable, and accommodated to the complexity of individual needs and schedules. Public transportation, cycling and pedestrian traffic receive nowhere near the type of government largesse and attention that the automobile does (see Litman's chapter). Quite apart from what people might feel in and about their cars, the extent of state support for policies and infrastructure that enable viable alternatives to car travel is one of the clearest factors that determine the degree to which people will feel compelled to drive.

Outside of the formal political realm, prevailing patterns of consumption are also reproduced by arbitrary work routines and scheduling constraints that structure our time and energies in ways that heighten our dependence upon the automobile and other consumer goods favouring speed and convenience. Currently, in Canada and most other industrial nations, the largest single segment of personal driving time is devoted to getting to and from work, followed by the time devoted to personal errands (Turcotte 2006). The growing distances between work, home, and other amenities, along with related issues of gridlock and congestion, mean that more time, fuel, and pollution is being expended in meeting these vital daily survival needs. Indeed, the increasing amount of time people are spending in their traffic has itself fed into the vogue for gas-guzzling, behemoth utility vehicles that can serve as placid, climate-controlled, and technologically decked-out 'carcoons' (Lyons and Urry 2005). In spite of the growing amount of time commuters are spending in their cars, the rapidity and convenience of automobile travel continue to significantly outpace that of public transportation and other options. In 2005, Statistics Canada reports (Turcotte 2006), 55 percent of workers travelling by car made the round trip between home and work in less than 60 minutes, whereas the portion of workers commuting by bus or subway doing so was only 13 percent. All told, the average round-trip of public transit users was 106 minutes, nearly double the 59 minutes spent by car commuters.

The steady growth of low-density suburban housing, along with acceleration of employment opportunities in peripheral urban areas under-serviced by public transit, has increasingly created complex commuting patterns that reinforce automobile dependency. To this extent, practical calculations of time efficiency and everyday convenience help us to understand why approximately 86 percent of Canadian workers currently use a car for all or part of the round-trip between home and work (Turcotte 2006). They also help us to understand why the automobile has become the preferred mode of travel for those – disproportionately women – charged with the responsibility of meeting domestic responsibilities outside of paid work – child and elder care, shopping, bill payments, health appointments and so on – while juggling competing time pressures and navigating between a number of disjointed destinations. Simple activities such as picking one's children up after work, taking one's parent to the doctor, or stopping at the store to get provisions for supper, become much more complicated, time-consuming and expensive when they involve multiple transit trips along irregularly scheduled routes whose stops may still be a considerable distance from one's intended destination (Turcotte 2006; Blumenberg and Waller 2003). Many features of contemporary life have increasingly cemented this relationship between driving, parenting and meeting other domestic responsibilities, increasingly privatizing everyday life patterns and pre-empting the development of more communal means of addressing our shared needs.

As such examples suggest, many of the worst consequences of automobile transportation today arise not from the frivolous nature of consumer wants, but from the ongoing relevance of legitimate human needs that, under prevailing living conditions, can only be met in unnecessarily wasteful, destructive and individualized ways. Within such constraining conditions, of course, cultural and ideological influences can play an important role in determining the specific ways in which people adapt to automobile-dependent lifestyles, and in increasing the relative social and environmental fallout of automobility. The historic surge in SUV sales, for instance, which has begun to reverse in an era of accelerating fuel prices and anti-SUV cultural backlash, was not simply a structural reflex of auto-dependent social conditions, but a reflection of other mediating influences related to processes of social class distinction, nostalgia for the 'off road' natural world, the security and privacy concerns of gridlocked commuters, prevailing cultural constructs of middle-class mothers as diligent family chauffeurs and so on (for further discussion, see for example, chapters by Conley and McLean).

While the variability of car culture needs to be acknowledged, understanding the roots of automobile- dependency requires us to move beyond the plane of individual motivations and desires, and look more closely at the decisions made by political and economic elites that have shaped the sphere of consumption. There are clearly a number of important vested interests underlying automobile dependent transportation systems, including not only the automobile industry itself, but petroleum and tire companies, road builders, real estate developers, construction companies and so on (see Henderson's chapter for a case study of conflicting vested interests). In many ways, such systems reflect the cosy and often incestuous relationship between private industry and government that prevails in much of the world today, which has tended to promote the ceaseless expansion of profit-yielding production and consumption regardless of its collective costs.

Viewed at the macro-level, automobile dependency and other forms of unsustainable consumption are not so much the pathological consequence of irresponsible personal behaviours as the predictable outcome of an economic model geared towards perpetual, profit-driven economic expansion, and the continual stimulation of private demand in the form of new needs and desires. Indeed, as Freund and Martin (1993, 5–6) insightfully argued:

> The widespread use of the private auto and its hegemony over other forms of transport manifest in a concrete form the major structural contradiction that haunts capitalist society, a contradiction first outlined by Karl Marx. Marx's form of the contradiction existed in the realm of production; essentially it was a contradiction between what is rational activity for individual capitalist firms (in their pursuit of profit) and what is rational for the economy and society as a whole. Auto hegemony represents a similar contradiction in the realm of consumption. Millions of individual drivers pursuing their rational self-interest in using autos for journeys to work, to shop, and to play create problems of exaggerated energy consumption, traffic congestion, and environmental degradation in the collective level – the level of the society and the economy.

In order to sustain the elevated levels of consumption required to absorb its ever-growing output, capitalism relies not simply on the 'false needs' created by advertising, but upon the real needs materialized within a social environment in which people have been made progressively more practically dependent upon market commodities. This privatization of survival was powerfully symbolized by hurricane Katrina in New Orleans, where those without access to a car were left to weather the storm alone, without socially organized transportation alternatives or adequate public provisioning of shelter, food and physical safety.

Conclusion

By highlighting the 'compulsory' or 'locked-in' quality of automobile dependency and other forms of consumer behaviour that are influenced by factors beyond the individual's immediate control, I certainly do not wish to advocate cynicism or passivity. Individual consumers can make many discretionary choices to lessen society's auto dependency and mitigate its negative impacts – from voluntarily reducing or foregoing car travel whenever possible, to choosing a place of residence that is close to work and is pedestrian and transit friendly, car-pooling, using the most fuel-efficient vehicle compatible with one's needs and so on. That said, such voluntary efforts soon bump up against the constraints of the social structures in which we live. In this sense, as Maniates (2002, 51) has argued, we need to address our shared problems not as conscientious consumers, but as "citizens who might come together and develop political clout sufficient to alter institutional arrangements that drive a pervasive consumerism." Ultimately, addressing auto dependency as a political problem – one that is intimately linked with prevailing social inequalities, systems of production, government policies, work patterns, time routines, gendered practices, social and material infrastructures, and so on – will require us to reclaim the value of collective action in pursuit of systemic social change. In the absence of this kind of broader political perspective, individuals will continue to experience automobile dependency as an unavoidable fact of nature, and the pursuit of alternatives will tend to get unduly narrowed to making greener consumer choices such as driving hybrid or alternative fuel vehicles.

By sanctifying the market as the supreme arbiter of societal decision-making, neoliberalism – a specific political and economic ideology operating at national and international levels – has restricted our democratic political possibilities (Bourdieu 1998; MacEwan 1999). It has reinforced the privatistic notion that all problems can and must be resolved by isolated individuals acting as rational economic agents (see also, in this volume, Trumper and Tomic's critique of neoliberalism). Confronting automobile dependency, however, highlights the limits of individualized solutions and points to the stubborn persistence of collective needs and aspirations that the market cannot satisfy. The market can respond to effective demand for a particular brand, colour, or make of car, but it cannot adequately respond to needs that cannot be easily expressed in commodity form: clean air and water; less cluttered and stressful personal schedules; safe, sociable and aesthetically pleasing neighbourhoods; access to public services; connection to nature; unstructured time to spend with friends and family; meaningful opportunities for participating in the decisions shaping the world in which we live. Auto-dependent suburbs themselves are, in some ways, the market's distorted response to such popular desires for safety, escape, conviviality, and connection to nature, but such solutions are necessarily intertwined with the very problems they are intended to solve.

Weaning ourselves from auto dependency will require a change in prevailing cultural values, although such values are not always the primary driver of the problems we face, and they are often more complex and ambivalent than we might think – a point that Conley (this volume), from a different angle, echoes in his analysis of the cultural codes of automobile advertising. Our attachment to consumer goods such as cars is often crosscut with strong feelings of frustration toward the everyday world in which they are enmeshed, and with inarticulate longings for a life in which they do not loom so large. Anti-car activism can take a variety of forms – from painting guerrilla bike-lanes on city streets, to participating in Critical Mass rides, to moving into the formal political arena to resist the expansion of the automobile infrastructure, champion the rights of non-drivers, and fight for radically new approaches to land use and transportation policy. All told, the critique of automobility is most promising when it moves beyond simply reorienting our personal relationship to a particular object, and begins to sharpen our resistance to the broader socio-econmoic system that conditions and presupposes the car's use. To this extent, the critique of automobility can become an entry-point into envisioning a post-consumerist future in which the expression of our needs is less distorted by capitalist imperatives, and our everyday lives are more democratically organized, personally fulfilling and attuned to natural limits.

References

Alvord, Katie. 2000. *Divorce your car! Ending the love affair with the automobile.* Gabriola Island: New Society Publishers.

Blumenberg, Evelyn and Waller, Margy. 2003. *The long journey to work: A federal transportation policy for working families.* Washington, DC: The Brookings Institution.   http://brookings.edu/es/urban/publications/20030801_Waller.pdf (accessed 25 May 2007).

Bourdieu, Pierre. 1998. *Acts of resistance: Against the tyranny of the market.* Richard Nice, trans. New York: The New Press.

Brower, Michael and Leon, Warren. 1999. *The consumer's guide to effective environmental choices: Practical advice from the union of concerned scientists.* New York: Three Rivers Press.

Durning, Alan. 1992. *How much is enough? The consumer society and the future of the earth.* New York: W.W. Norton and Company.

Freund, Peter and Martin, George. 1993. *The ecology of the automobile.* Montreal: Black Rose Books.

Gardner, Gary, Assadourian, Erik and Sarin, Radhika. 2004. The state of consumption today. In *State of the world 2004 – Special focus: The consumer society*, Linda Starke, ed., pp. 3–21. New York: W.W. Norton and Company.

Handy, Susan. 2006. The road less driven. *Journal of the American Planning Association* 72(3): 274–78.

Hertsgaard, Mark. 1998. *Earth odyssey: Around the world in search of our environmental future.* New York: Broadway Books.

Kunstler, James Howard. 1998. *Home from nowhere. Remaking our everyday world for the twenty-first century.* New York: Simon and Schuster.

Litman, Todd. 2001. The costs of automobile dependency and the benefits of transportation diversity. Victoria, BC: Victoria Transport Policy Institute. http://www.vtpi.org (accessed 25 May 2007).

Lodziak, Conrad. 2000. On explaining consumption. *Capital and Class* 72: 111–33.

——. 2002. *The myth of consumerism.* London: Pluto Press.

Lyons, Glenn and Urry, John. 2005. Travel time use in the information age. *Transportation Research Part A* 39(2–3): 257–76.

MacEwan, Arthur. 1999. *Neoliberalism or democracy? Economic strategy, markets, and alternatives for the 21st century.* London: Zed Books.

Maniates, Michael. 2002. Individualization: Plant a tree, buy a bike, save the world? In *Confronting consumption*, Thomas Princen, Michael F. Maniates and Ken Conca, eds, pp. 43–66. Cambridge: The MIT Press.

Newman, Peter and Kenworthy, Jeffrey. 1999. *Sustainability and cities: Overcoming automobile dependence.* Washington: Island Press.

Nielsen, Ron. 2006. *The little green handbook: Seven trends shaping the future of our planet.* New York: Picador.

Princen, Thomas. 2005. *The logic of sufficiency.* Cambridge: The MIT Press.

Renner, Michael. 2005. Vehicle production sets new record. In *Vital signs 2005: The trends that are shaping our future*, Linda Starke, ed., pp. 56–57. New York: W.W. Norton and Company.

——. 2006. Vehicle production continues to expand. In *Vital signs 2006–2007: The trends that are shaping our future,* Linda Starke, ed., pp. 64–65. New York: W.W. Norton and Company.

Sanne, Christer. 2002. The willing consumers – or locked-in? Policies for a sustainable consumption. *Ecological Economics* 42: 273–87.

——. 2005. The consumption of our discontent. *Business Strategy and the Environment* 14: 315–23.

Sawin, Janet. 2004. Making better energy choices. In *State of the world 2004 – Special focus: The consumer society*, Linda Starke, ed., pp. 24–43. New York: W.W. Norton and Company.

Sheehan, Molly O'Meara. 2001. *City limits: Putting the brakes on sprawl.* Washington: Worldwatch Institute.

Sheller, Mimi. 2004. Automotive emotions: Feeling the car. *Theory, Culture and Society* 21(4/5): 221–42.

Simms, Andrew. 2005. *Ecological debt: The health of the planet and the wealth of nations.* Ann Arbor: Pluto Press.

Statistics Canada. 2006. *Human activity and the environment: Annual statistics 2006.* Ottawa: Statistics Canada, Environment Accounts and Statistics Division, Catalogue no. 16-201-XPE.

Turcotte, Martin. 2006. *The time it takes to get to work and back, 2005.* General Social Survey on Time Use: Cycle 19. Ottawa: Statistics Canada.

UNDP (United Nations Development Programme). 1998. *Human development report 1998.* New York: Oxford University Press.

Urry, John. 2004. The 'system' of automobility. *Theory, Culture and Society* 21(4/5): 25–39.

US Department of Transportation, Federal Highway Administration. 1997. *Our nation's travel: 1995 NPTS early results report.* Washington, DC: FHWA.

Wachtel, Paul L. 1983. *The poverty of affluence: A psychological portrait of the American way of life.* New York: The Free Press.

PART 4

Beyond the Car

Chapter 11

Mobility as a Positional Good: Implications for Transport Policy and Planning

Todd Litman

Contentment is natural wealth, luxury, artificial poverty
(Socrates 469–399 BC).

The popular refrain that Americans (or the British, French, or other groups) have a love affair with their cars attempts to express the emotional as well as practical relationships between consumers and motor vehicles in modern societies. But such claims fail to accurately express the true relationship between consumers and automobiles. Other chapters in this volume have called into question this understanding of automobility, which can also be compared with a marriage of convenience or even a 'shotgun' (forced) marriage. Soron argues that the spatial organization of cities makes automobile consumption compulsory, and Dennis and Urry likewise argue that once automobility is locked in, cars are more convenient than other forms of transportation; in contrast chapters by Conley, McLean and Simons focus on cultural representations of speed, excitement, status, domination, and sacrifice as contexts for the consumption of automobiles.

This chapter examines the role that social status plays in favouring automobile ownership and use, resulting in excessive automobile dependency at both individual and community levels. It investigates the degree to which motorized travel is a 'prestige' or 'positional' good – that is, a good that people often consume to enhance their status. Although economists have discussed the general and theoretical implications of positional goods for decades, their research on mobility as a positional good is relatively limited, suggesting that it is fertile ground for analysis and application to decision-making.

When evaluating consumer behaviour and benefits, economists sometimes distinguish between goods' functional value (which provides direct utility) and their positional or prestige value which is intended to enhance the consumers' social status (Hirsch 1976). Examples of positional goods include fashionable jewellery and clothing, ostentatious homes, luxurious vehicles, and extravagant entertainment. Consumption of these goods is considered 'conspicuous', meaning that it is intended to be noticed and valued by other people (Veblen 1899). Conceptual tests of positional value are, 'Would I choose this particular good if it were unpopular?' and 'Would I choose this good if nobody else knew?' Prestige value is often a component of functional goods. For example, many motorists choose vehicles with greater potential speeds and off-road abilities than actually needed because they believe these features have prestige.

From an individual's perspective positional goods can provide significant functional benefits by enhancing a person's status, and therefore their social standing in a community and their economic opportunities. Having a prestigious vehicle can increase a young person's chance of dating, and therefore, mating a popular partner. Employees can enhance their self confidence and careers by driving a fashionable car. Living in a prestigious neighbourhood raises a person's social status and networking opportunities. Business competitiveness often requires accommodating customers' preferences for status goods. An offhand judgment about a person's transportation ('Take me away from here in that nice car of yours', or 'He's riding a loser cruiser') can cause delight or injury. Popular culture embodies vehicles and travel decisions with symbolic value – they help define a person's identity (as illustrated in Genovese's chapter).

However, from society's overall perspective, positional goods provide little or no net benefit because gains to one individual are offset by losses to others (Hirsch 1976; Frank 1999). For example, if one person drives a prestigious car his or her peers must obtain equally prestigious vehicles to maintain status. It represents a form of inflation, popularly called 'keeping up with the Joneses', that raises everybody's costs without increasing overall welfare. Positional value is therefore an economic trap, a situation in which individuals compete in ways that waste resources (also called a social trap, reflecting society's overall perspective, a zero sum game, reflecting the fact that gains to one represent losses to another, or a treadmill, because to the degree that social position is based on economic success in competitive conditions, people feel that they must work harder to maintain a given level of status). Described differently, prestige value is an economic transfer rather than a net economic gain.

This chapter investigates how positional value affects transportation decisions, explores the resulting economic impacts (including impacts on social welfare and external costs), and discusses implications for transport policy and planning. By drawing on a social welfare perspective, this chapter argues that – on both efficiency and equity grounds – policies to reduce wasteful and non-sustainable transportation consumption can be justified.

The Science of Happiness

When evaluating economic progress people often use indicators of material wealth and productivity such as changes in income, property ownership, and Gross Domestic Product, assuming that increased wealth increases happiness (Redefining Progress 2006). But material wealth is just one factor affecting happiness (which economists call social welfare; in practice, they often ignore that critical concept). To evaluate the overall value of positional goods requires a deeper understanding of how consumption decisions affect happiness.

Developed countries have achieved a high level of material wealth that could provide a high level of social welfare. But happiness is elusive. Residents of wealthy countries complain about excessive stress, inadequate leisure time and social isolation (Easterbrook 2003). If we had twenty-first century productivity with nineteenth century expectations we would live in Eden, but economic traps erode much of the potential welfare gains from material progress, reducing the efficiency with which we achieve happiness.

Researchers have investigated factors that affect how much happiness people achieve and the efficiency with which wealth provides happiness (Frank 1999; Stutz 2006; Dolan et al. 2006). This research indicates that a rise from poverty to moderate wealth significantly increases happiness, but once people's basic material needs for food, housing and medical care are met, the relationship between wealth and happiness varies widely. If used efficiently, more wealth can increase happiness, but if squandered, it may provide little additional happiness. Some people learn to be happier with less wealth, as illustrated in Figure 11.1, for example, by choosing a more enjoyable but less lucrative job, or by retiring.

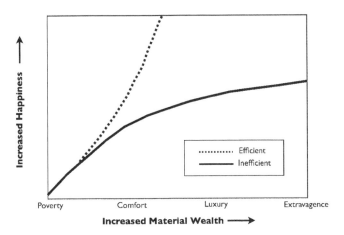

Figure 11.1   Wealth and Happiness. Based on Stutz 2006

Source: Based On Stutz 2006.

Economic traps reduce the efficiency with which wealth creates happiness (Easterbrook 2003). As a community becomes wealthier, the quantity of consumption required to achieve a given level of status, and the portion of consumption devoted to positional value tend to increase, as illustrated in Figure 11.2.

In addition, material affluence (abundant money) often requires sacrificing time affluence (abundant free time) and social affluence (abundant friendships). Since material affluence is more conspicuous than other types of affluence, positional competition skews people's decisions toward more employment and consumption

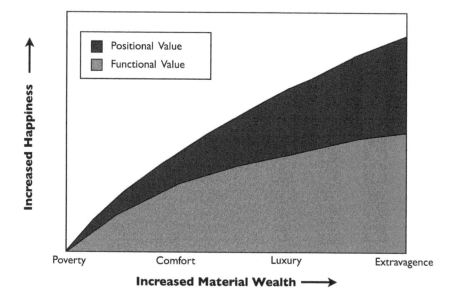

Figure 11.2   Positional and Functional Value. Based on Galbraith 1958; Stutz 2006

Source: Based on Galbraith 1958; Stutz 2006.

than optimal, that is, beyond what people would choose if they expressed their true preferences rather than social expectations. For example, it dissuades people from choosing lower-paying but more satisfying jobs, or working fewer hours to have more time to spend with family and friends because such a lifestyle has less prestige. As a result, people often work long hours to afford expensive holidays they feel are needed to recover from job stress. Increased consumption may also contribute to social and health problems such as obesity, smoking, alcoholism and drug use. The contradictory effects of positional consumption help to explain why happiness is so elusive (Scitovsky 1976).

Starting from poverty, increased material wealth provides significant benefits (happiness) by improving health and comfort, but once people's basic physical needs are met, an increasing portion of wealth is devoted to positional goods. These goods raise the status of people who consume them but reduce the status of others and so provide no net social benefit.

Both individual and public decisions affect the efficiency with which wealth provides happiness. Individuals can choose lower paying but more satisfying jobs, avoid wasting money on goods that provide little real enjoyment, and choose friends who value their personality rather than their wealth. But some economic traps of affluence reflect public policies and community values. For example, many people might be happier overall using cheaper travel modes such as walking, cycling and public transit, provided they are convenient, safe and socially acceptable. If planning decisions or social attitudes favour automobile travel, people will be forced to drive more than they actually prefer (what Soron, this volume, calls 'compulsory consumption').

For these reasons, consumption of positional goods reduce the net welfare benefits resulting from consumer expenditures, a significant factor that economists and decision-makers should consider when evaluating policies (Stutz 2006). A welfare economic perspective, which examines the relative needs that consumption satisfies as a form of consumption externalities, brings such a consideration into analysis (Verhoef and van Wee 2000). To the degree that positional value stimulates consumption of resources, such as energy, minerals and land, it contradicts sustainability objectives, such as cultural and ecological preservation (TRB 1997; Litman and Burwell 2006).

Because sustainability requires constraining material consumption within ecological limits (such as limiting consumption of land in order to protect habitat, and reducing fossil fuel consumption to minimize climate change), sustainable development requires maximizing the efficiency with which material wealth provides happiness, as indicated by dashed arrows (see Figure 11.3). However, economic traps tend to shift consumers toward less efficient and therefore less sustainable resource consumption, as indicated by solid arrows.

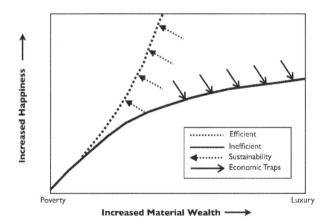

Figure 11.3   Sustainable Development

Many economists and public officials are understandably reluctant to support strategies that contradict consumer sovereignty (the idea that consumers should be free to choose the goods they prefer). It may, however, be rational and beneficial for policymakers to implement policies such as special taxes on luxury goods and more progressive tax structures that avoid these economic traps (Frank 2005). Society often intervenes in markets to reduce unjustified harm, such as outlawing drugs, and to favour basic services over luxury services, such as subsidizing health care. Policies to reduce the harm of prestige-induced wasteful consumption can be justified on both efficiency and equity grounds.

Transportation Impacts

Economists have given little attention to understanding the ways that prestige value impacts mobility. Positional value influences transportation activities in several ways: it increases motor vehicle ownership and use; it leads to ownership of more expensive vehicles; it stigmatizes alternative modes of travel; it contributes to urban sprawl; it promotes long-distance recreation travel; it shapes public policies that contribute to auto dependence; and it leads to over-capitalization in the auto industry. It supports numerous decisions by individual households, businesses and governments that favour automobile travel, reduce travel options and stimulate automobile-oriented land use patterns, leading to compulsory consumption as discussed in Soron's chapter.

Motor Vehicle Ownership

Positional value motivates some people to increase their vehicle ownership beyond what they would otherwise choose (Steg 2005). For example, a lower-income person might be best off overall relying on a combination of walking, cycling, public transit, and rented cars, but chooses instead to own an automobile because of the status it conveys. Many teenagers take part-time jobs in order to afford a vehicle that they use primarily to commute to work, a pattern that is only rational if the car ownership is an end in itself. Though it would be cheaper to rely on taxis, some seniors who seldom drive own vehicles because not driving is stigmatized (Adler and Rottunda 2006).

   Once people own vehicles they are motivated to use them in order to maximize the value of their fixed expenses. The prestige value of vehicle ownership therefore shifts people from multi-modal transportation (in which they share vehicles and use various modes) to automobile dependency (each driver has a personal vehicle to use for most travel). In automobile-dependent societies, automobiles become important status symbols that many people use to display their identity. People do not own an automobile are considered 'nobodies'. In addition, non-drivers frequently face practical problems, such as inadequate walking and cycling conditions, poor public transit service, and unpleasant stops and station waiting areas. As a result, consumption of cars becomes increasingly compulsory (Soron, this volume).

   It is difficult to determine the magnitude of the positional good effect. Motor vehicles provide significant functional benefits (see Dennis and Urry's chapter on the convenience of the automobile); prestige value alone increases vehicle ownership only modestly, perhaps 5–15 percent in the short term, reflecting marginal value automobiles, such as vehicles owned by lower-income residents of communities with good transit service (who could manage without an automobile) and households' second, third, or fourth car. These impacts probably increase over the long run as higher vehicle ownership further increases automobile dependency, as described below.

Luxury Status Vehicles

Prestige value motivates many consumers to purchase more expensive vehicles than they otherwise would (Carlsson et al. 2003). This propulsion is illustrated by the prominence luxury vehicles receive in status-oriented publications such as the *Robb Report* and *Millionaire Magazine*. Such vehicles may provide some functional benefits compared with cheaper and more practical vehicles, such as increased reliability, durability, and safety (although not always, e.g., sports cars; see also MacGregor and Wetmore, both in this volume), and more pleasurable driving (although not always, e.g., large SUVs), but much of their attraction is positional. The automobile industry often highlights the status value of luxury vehicles in marketing, as described in chapters by Conley and McLean.

Only a small portion (perhaps 10 percent) of the overall fleet can be considered truly luxury vehicles, but positional value encourages consumers to choose higher value vehicles than they would if such vehicles lacked status value or did not embody symbols of group identity. For example, some groups value 4 x 4 trucks and SUVs, while others value low-riders built for cruising, or sports cars designed for performance. Embodying automobiles with status increases vehicle ownership, vehicle costs, and vehicle travel as owners drive more to display their vehicles and justify their investments. Many vehicle prestige features (larger size, increased performance, off-road capability, additional accessories) reduce fuel efficiency, leading to increased resource consumption that provides little net social benefit (Verhoef and van Wee 2000).

Mode Choice

As more and more countries develop an automobility culture (Urry 2004; see also Martin, as well as Trumper and Tomic, this volume), automobile travel has gained prestige and alternative modes such as walking, cycling and public transit have lost prestige, and indeed, are often stigmatized (Ory and Mokhtarian 2005). A survey of Dutch commuters found that their decision to drive rather than use other modes resulted more from symbolic than from functional motives (Steg 2005). The stigma of alternative modes is illustrated by the popular term 'loser cruiser' – defined by the *Encarta Dictionary* as "transportation for people without vehicles: a public transportation vehicle such as a bus or minivan, used by, e.g. college students or service members without personal vehicles" (MSN 2007) – which stigmatizes use of public transportation and insults people who do not drive. This stigmatization reduces use of alternative modes compared with what consumers would otherwise choose. For example, commuters are more likely to drive rather than use public transit, and parents are more likely to chauffeur children rather than allow them to walk or bus to school (Steg 2005).

The stigmatization of walking, cycling and public transit travel also has indirect effects. These modes experience significant economies of scale; reductions in their demand reduce their quality of service, and reduce the incentive for multi-modal land use patterns. For example, if driving has more prestige than other modes, businesses will locate to maximize access by automobile rather than other modes, as described in the following section. Another indirect effect of stigmatized mobility modes is that people feel obliged to work more in order to afford car ownership, for the sake of prestige. Because prestige travel modes tend to be faster but more costly than stigmatized modes, a person might work longer hours despite their preference to work fewer hours (Tranter 2004).

The magnitude of this impact of stigmatized travel is indicated by the much higher automobile mode split in North American cities compared with equally wealthy European cities that retain respect and social support for walking, cycling and transit travel, as indicated in Figure 11.4. Of course, factors such as quality of service and land use patterns also affect travel behaviour, but these result, in part,

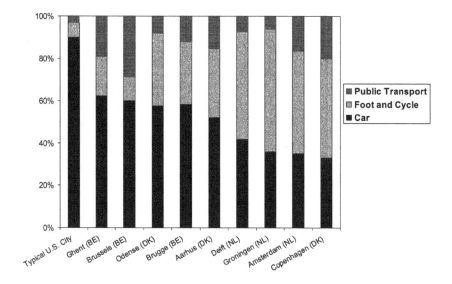

Figure 11.4   Mode Split in Selected European Cities. Based on ADONIS 1998

Source: Based on ADONIS 1998.

from long-term decisions made by individual households (such as which home to purchase) and communities (such as transportation planning decisions and land use policies) which reflect and reinforce the prestige of each mode.

Although it is difficult to determine exactly how much the stigmatization of alternative modes affects travel behaviour, the total impact is probably moderate to large, particularly in urban areas over the long run, increasing automobile travel 10–20 percent more than would otherwise occur.

Automobile-Dependent Land Development (Sprawl)

In North America, single-family homes and suburban locations generally have more prestige than more compact homes in urban neighbourhoods. Market surveys indicate that many households choose suburban homes primarily for their social attributes (prestige, security, and good schools) rather than physical attributes such as large lawns (NAR and NAHB 2002; Litman 2007). For example, in the film *American Beauty* (1999), the mother, frustrated that her teenage daughter does not appreciate the family's material wealth, as indicated by their perfectly-maintained suburban home, proclaims, "When I was your age, I lived in a duplex!" This comment plays on the popular idea that anything other than a single-family suburban home is socially inferior.

The positional value of single-family, suburban housing tends to shift housing location decisions toward more automobile-dependent neighbourhoods than what consumers would consider optimal based only on their physical needs and preferences, as illustrated in Figure 11.5. For example, a household that would otherwise prefer an apartment shifts to a townhouse, a household that would otherwise prefer a townhouse shifts to a small-lot single-family house, and a household that would otherwise prefer a small-lot single-family house chooses a large-lot house, because they have more prestige (Levine and Frank 2007).

People who live and work in suburban locations tend to drive significantly more than residents of more multi-modal urban locations. If more multi-modal, urban locations were as prestigious as suburban locations, perhaps 10–30 percent of suburban households would shift to such locations, reducing their per capita vehicle miles of travel 20–40 percent, providing total reductions of 2–11 percent.

Longer Distant Recreation Travel

Positional value also promotes demand for long-distance holiday trips, especially those associated with exotic destinations (Veblen 1899; Duffy 2002; Ory and Mokhtarian 2005). As international travel becomes more common, more exotic destinations are needed for a trip to have unusual value, causing vacationers to travel greater distances than they functionally enjoy. As a result, many tourists travel to distant resorts with rushed schedules that include little or no interaction with local people or culture (many resorts are designed to prohibit such interactions), and while there demand services (food, shelter, and entertainment) identical to what

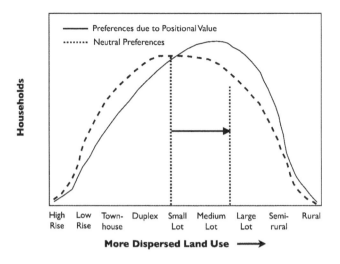

Figure 11.5   Suburban Location Prestige Shifts Consumer Decisions

they could obtain closer to home. Often, many four-day visits require two days of travel (one day each way), so travelers have little time to relax and enjoy the experience. Much of the pleasure of such trips comes from the sense of having visited a distant location, and the resulting bragging rights.

Because people use such travel to compete for prestige, demand for this type of travel is virtually unlimited. If international travel were sufficiently cheap, parents might have birthday parties in distant lands, even for children too young to appreciate the experience, simply to make it a 'special' event. If interplanetary travel were sufficiently cheap, an earth-bound holiday might be considered dull. It is difficult to determine how much positional value affects total travel. Many people are sincerely interested in visiting distant and unusual destinations, but the prestige of such trips probably increases some North American travel to exotic locations, such as Cancun, Mexico, and Tahiti.

Planning Practices

For much of the last century, transportation and land use planning practices tended to favour automobile and air travel over other modes, with dedicated funding, minimum parking requirements, and transport system quality indicators that primarily considered automobile travel conditions. These practices partly reflect the sense by public officials that these modes are prestigious. This effect is often subtle, reflected in extra enthusiasm for automobile improvements and weaker support for efforts to improve alternative modes. These practices are particularly obvious in developing countries where the majority of transport resources are often devoted to improving automobile transportation, although the majority of residents rely on other modes (see Trumper and Tomic's chapter on Chile). Although it is difficult to determine exactly how much prestige value in land planning contributes to automobile dependency, even a small contribution could have large effects on travel over the long term.

Industrial Development Policy

Many people, including decision-makers, consider the vehicle manufacturing industry particularly important to the economy, and a glamorous symbol of economic and technical progress, leading to public policies that favour it over other sectors. The close tie of the automobile to the 'glamour' of modernity has been a motivating force not only in developed, but increasingly, in developing countries (see Martin, and also Trumper and Tomic, this volume). Many jurisdictions offer generous public subsidies to attract vehicle production facilities, often much greater per job than offered to other industries and exceeding what is economically justified. The global vehicle industry is overcapitalized, with significant excess capacity and numerous countries competing to expand their exports, resulting in low return on investment for most producers (PricewaterhouseCoopers 2007). Most countries would be better off supporting other industries that provide greater returns and more competitive advantage. While it is difficult to know how much perceptions of the glamour and prestige of the motor vehicle industry affects travel behaviour, it has probably increased vehicle production and therefore vehicle ownership and use than is economically optimal.

Summary

Table 11.1 summarizes the positional value categories identified in this chapter and their transport impacts. Although these impacts may individually seem modest, stimulating vehicle travel just a few percent, their effects are cumulative. Their total travel impacts are probably moderate to large, increasing per capita vehicle ownership 5–15 percent and vehicle travel 10–30 percent in the short term, and more over the long term, compared with what would occur if motor vehicle travel were not considered prestigious and alternative modes were not stigmatized.

This conclusion is supported by the significantly higher rates of automobile travel in communities where walking and transit are stigmatized compared with equally wealthy communities where walking and transit travel are respected. For example, in most US cities 80–90 percent of trips are made by automobile, compared with 40–60 percent of trips in wealthy European cities (ADONIS 1998). Although it is difficult to separate out the specific factors that affect travel, such as fuel prices and transit service quality, the relative levels of prestige of different modes by public officials and consumers is undoubtedly an important factor.

Table 11.1   Summary of Positional Value Travel Impacts

Category	Description	Travel Impacts
Vehicle Ownership	Households own more vehicles than functionally justified or cost effective.	Increases vehicle ownership and therefore use.
Luxury Vehicles	Motorists choose more valuable vehicles.	Increases vehicle costs. Stimulates some additional vehicle travel.
Mode Choice	Alternative modes (such as walking, cycling, ride-sharing, and public transit) are stigmatized relative to driving.	Reduces use of alternative modes, and over the long term, reduces their quantity and quality, increasing automobile travel.
Long-distance Recreational Travel	Consumers choose more distant holiday travel destinations.	
Planning Practices	Planners and public officials favour automobile and air travel.	Increases investment in automobile and air travel, and reduces the quality of alternatives.
Industrial Policy	Public officials favour motor vehicle industries more than economically justified.	Increases automobile ownership and use.

Source: Litman 2006.

Travel demand has a long tail, meaning that if the price (perceived user costs) declines, consumers will increase their mobility, as illustrated in Figure 11.6, in part due to competition for status. For example, if financial and time costs were low enough, Los Angeles residents would travel to New York for dinner, London for a show, and return home to sleep in their own bed, in part so they can brag about their worldliness, and avoid embarrassment if their neighbours brag about making such trips. The additional travel provides minimal user benefit, because it consists of travel that consumers will only take if their costs are low enough and forego if their costs increase.

The demand curve for mobility has a long tail: as prices decline mileage increases even if the additional travel provides small incremental benefits and imposes significant external costs. As a result, an increasing portion of travel has negative social value (total benefits are less than total costs, including energy and environmental externalities), indicated by the shaded area in Figure 11.6. This increase in vehicle travel imposes various costs on society and increases various transportation problems (see Table 11.2).

Table 11.2: Summary of Costs of Vehicle Travel

Traffic congestion	Dispersed urban-fringe development (sprawl)
Road and parking facility costs	Reduced mobility options for non-drivers
Traffic accidents	Reduced fitness and health
Energy consumption	Tourism impacts on traditional societies and natural features
Pollution emission	

Source: Litman 2006.

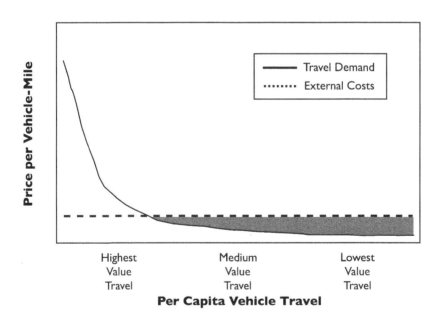

Figure 11.6   Travel Demand Curve

As a result, much of this additional travel is economically inefficient. In addition, its positional value makes increased, motorized transport inequitable, reducing the status of physically, economically, and socially disadvantaged people who rely on lower status modes (see Martin, this volume, for a discussion of disadvantaged access to transportation). In particular, it reduces the quality of walking, cycling and public transit transport, and stigmatizes them, further reducing the status of disadvantaged people.

Described more positively, increasing the status of efficient transportation options such as walking, cycling, ride-sharing and public transit can help reduce problems such as traffic congestion, accidents, energy dependence and transportation inequity. This suggests that marketing strategies that destigmatize alternative modes may be as important as engineering improvements in solving future transportation problems (see Dennis and Urry's chapter for an examination of both).

Possible Offsetting Benefits

It could be argued that positional value provides social benefits that offset their costs. Prestige value motivates consumers to earn more money, and so stimulates education, employment and economic productivity. However, this is not unique to transport and there is little evidence that transport prestige value is better at motivating productivity gains than other goods, such as attractive homes and new electronics. To the degree that prestige mobility imposes large external costs (traffic congestion, road and parking facility costs, accidents, pollution, habitat loss and so on), it is economically inefficient to encourage such consumption patterns.

Prestige value stimulates technological innovation and industrial development and demand for prestige products encourages motor vehicle manufacturers to innovate. During certain time periods in certain regions, motor vehicle production helped stimulate industrial development, but the global vehicle industry is now experiencing overcapacity, making vehicle manufacturing relatively unprofitable while other industries, such as electronics and software development, now provide greater innovation returns on investments (*The Economist* 1999).

Prestige value also stimulates total resource consumption. To the degree that a healthy economy needs overall consumption to be stimulated (an assumption of Keynesian economics), prestige-stimulated mobility can be considered an ideal strategy since it has almost unlimited potential to consume. However, since motor vehicle travel imposes significant external costs, increasing its consumption is less desirable than increasing consumption of other goods that have fewer external costs and greater external benefits. For example, society would probably be better off if households compete for status based on their education achievement, the beauty of their gardens, or their community contributions than the value of their vehicles.

It is possible that other forms of transport may become more prestigious, reducing or even reversing biases favouring motor vehicle travel and sprawl (but see Dennis and Urry, this volume, who argue that because of its convenience, the automobile is here to stay as the dominant and most valued form of transport). For example, some social groups consider bicycle commuting prestigious, and some communities (particularly large cities with well-established rail transit systems) have no stigma associated with transit travel. In recent years, popular culture, as expressed in films and television shows such as *Seinfeld* and *Sex in the City*, has glamorized urban lifestyles, at least for young professionals and retirees. However, these are exceptions, and are generally overwhelmed by biases favouring motorized transport. Much greater shifts will be needed before transport decisions could be considered unbiased, efficient and equitable.

Implications for Planning

Prestige value has significant implications for transport and land use planning. It reduces net benefits of some mobility activities (particularly automobile and air travel), increases transport problems and makes travel demand virtually unlimited. Although it is difficult to quantify the total effects of positional value on mobility, the direction of impacts is clear: it erodes the net welfare gain of increased vehicle use and increases external costs. Trying to satisfy the additional mobility demand stimulated by prestige value is economically inefficient and unfair to people who rely on alternative modes, and so are worse off with increased automobile dependency.

Good public policy favours necessities over luxuries (Kemp 1998), and so should favour basic mobility (transport activity considered socially valuable) over less important transport activity (VTPI 2006a). This policy approach need not require strict regulations or rationing; it may simply involve modest changes in planning and pricing practices to favour alternative modes and test consumer demand for luxury mobility (Litman 2005).

Increased motor vehicle ownership and mobility may provide indirect benefits by supporting particular industries and innovations. However, this undue regard for motor vehicle transport reduces development of other industries and their innovations. To the degree that consumers truly enjoy mobility-intensive goods and activities and are willing to pay their full costs, there is no social justification for constraining them. However, there is no reason for policies to favour mobility-intensive industries and goods over others, or to under price such goods relative to their full costs.

Below are specific examples of public policy strategies to reduce the negative effects of transport prestige value (Cairns et al. 2004; VTPI 2006b):

- Favour basic, functional mobility over luxury mobility in transport planning and pricing.
- Avoid letting prestige value bias planning decisions to favour automobile and air travel to the detriment of alternative modes such as walking, cycling, and public transit. Review current transport planning and funding practices

to identify and correct unintended biases.

- Apply luxury taxes, road pricing, and emission fees to internalize external costs and capture some of the value that would otherwise be lost through competition for status.
- Use positional value in marketing to help achieve transport planning objectives (VTPI 2006b). For example, it is important to use market techniques to raise the status of alternative modes (walking, cycling and public) to citizens and users. It may be appropriate to enhance the status of public transit by implementing urban rail rather than bus transit, and providing more attractive and comfortable stations and vehicles.
- Promote non-motorized modes as part of a desirable lifestyle.
- Promote walkable, mixed-use, transit-oriented communities as desirable places to live, and an automobile-dependent community as outdated and unsophisticated.

Conclusions

Positional goods confer status on their users, but this benefit is offset by reduced status to others, resulting in little or no net benefit to society. If the additional consumption stimulated by prestige value imposes external costs it can make society worse off overall. This economic trap wastes resources. As society becomes wealthier an increasing portion of consumption reflects positional value.

Positional value significantly affects transportation decisions. Although these impacts are difficult to quantify, their direction is clear: positional value tends to favour more costly, faster, resource-intensive modes at the expense of cheaper, slower, more resource-efficient modes, which increases per capita vehicle ownership and travel, reduces use of alternative modes, stimulates sprawl and encourages more distant holiday destinations. The total impacts are significant, probably increasing vehicle travel by 10–30 percent in the short term and more over the long term. Since motor vehicle travel and sprawl impose significant external costs, these impacts tend to harm society overall.

This impact has several implications for transport policy and planning decisions. It suggests that functional travel demand (the level of mobility needed to satisfy economic needs) is significantly lower than what is reflected by consumer behaviour; that a significant and growing portion of transport activity provides negative net social benefit; and that efforts to satisfy prestige-stimulated transport demand is both futile and economically inefficient because it tends to raise the level of mobility that consumers consider necessary to maintain their status. It is important that decision makers understand how prestige value affects travel behaviour, and the full economic, social and environmental costs that result, and take these impacts into account when evaluating policy and planning options.

Prestige value may provide indirect benefits, such as supporting certain industries and stimulating innovation. But existing evidence does not justify favouring such industries over others and does not suggest that consumption of mobility stimulates more innovation than other goods. Economics is concerned with maximizing social welfare, that is, total human happiness. Sustainability requires constraining resource consumption to accommodate long-term ecological and social limits. Sustainable economics therefore requires increasing the efficiency with which resource consumption provides humans happiness. Positional goods contradict this by reducing the social welfare benefits provided by each unit of resource consumption (such as a gallon of fuel or an hour of travel). Positional goods therefore tend to contradict sustainability objectives.

Although previous authors have examined the economic implications of positional goods, and some commonly-recognized positional goods involve mobility, little formal research has focused on mobility as a positional good. This appears to be a new research subject. More research is justified to better understand how prestige values affect travel behaviour, and how the status of mobility-related goods can be better aligned with society's long-term planning objectives.

References

Adler, Geri and Rottunda, Susan. 2006. Older adults' perspectives on driving cessation. *Journal of Aging Studies* 20(3) September: 227–35.

ADONIS. 1998. Best practice to promote cycling and walking. Analysis and development of new in-sight into substitution of short car trips by cycling and walking (ADONIS). European Commission, Fourth Framework Programme. Copenhagen, Denmark: Road Directorate. http://www.vejdirektoratet.dk/dokument.asp?page=document&objno=7134 (accessed 13 September 2006).

American Beauty. 1999. Alan Ball (writer) and Sam Mendes (director). Glendale, CA: DreamWorks.

Cairns, Sally, Sloman, Lynn, Newson, Carey, Anable, Jillian, Kirkbride, Alistair and Goodwin, Phil. 2004. Smarter choices – Changing the way we travel. Case study reports. London, UK: Department for Transport.

Carlsson, Fredrik, Johansson-Stenman, Olof and Martinsson, Peter. 2003. Do you enjoy having more than others? Survey evidence of positional goods. *Working Papers in Economics* 100 (May). http://www.handels.gu.se/epc/archive/00002855/01/gunwpe0100.pdf (accessed 13 September 2006).

Dolan, Paul, Peasgood, Tessa and White, Mathew. 2006. Review of research on the influences on personal well-being and application to policy making. London: UK Department for Environment, Food and Rural Affairs.

Duffy, Rosaleen. 2002. *A trip too far: Ecotourism, politics and exploitation.* London, UK: Earthscan James and James.

Easterbrook, Gregg. 2003. *The progress paradox: How life gets better while people feel worse.* New York: Random House.

The Economist. 1999. The car industry. *The Economist* 13 February: 23–25.

Frank, Robert. 1999. *Luxury fever: Why money fails to satisfy in an era of excess.* New York: Free Press.

———. 2005. Positional externalities cause large and preventable welfare losses. American Economic Association Annual Meeting, Expanding the Frontiers of Economics, Philadelphia, PA, 7–9 January. http://www.aeaweb.org/annual_mtg_papers/2005/0108_1015_0601.pdf (accessed 13 September 2006).

Galbraith, John Kenneth. 1958. *The affluent society.* Boston: New American Library.

Hirsch, Fred. 1976. *Social limits to growth.* Cambridge: Harvard University Press.

Kemp, Simon. 1998. Perceiving luxury and necessity. *Journal of Economic Psychology* 19: 591–606.

Levine, Jonathan and Frank, Lawrence D. 2007. Transportation and land-use preferences and residents' neighborhood choices: The sufficiency of compact development in the Atlanta Region. *Transportation* 34(2) March: 255–74.

Litman, Todd. 2005. Win-win transportation solutions: Cooperation for economic, social and environmental benefits. Victoria, BC: Victoria Transport Policy Institute.

———. 2006. Transportation cost and benefit analysis. Victoria, BC: Victoria Transport Policy Institute.

———. 2007. Where we want to be: Household location preferences and their implications for smart growth. Victoria, BC: Victoria Transport Policy Institute.

Litman, Todd and Burwell, David. 2006. Issues in sustainable transportation. *International Journal of Global Environmental Issues* 6(4): 331–47.

Millionaire Magazine Shopping and Entertainment. N.d. http://www.millionaire.com (accessed 13 September 2006).

MSN. 2007. 'Loser cruiser'. MSN, Microsoft: Encarta World English Dictionary. http://encarta.msn.com/dictionary_701707214/loser_cruiser.html (accessed 13 September 2006).

NAR and NAHB. 2002. Smart growth: Building better communities. Chicago: National Association of Realtors. http://www.realtor.org/SG3.nsf/Pages/NAR NHAB02Survey?OpenDocument (accessed 13 September 2006).

Ory, David T. and Mokhtarian, Patricia L. 2005. When is getting there half the fun? Modeling the liking for travel. *Elsevier Science* 39(2–3): 97–123.

PricewaterhouseCoopers. 2007. Global automotive financial review. PricewaterhouseCoopers' Automotive Practice. http://www.pwc.com/extweb/pwcpublications.nsf/docid/CBADDEEABACD24BC8525736E007CFD9B (accessed 8 July 2008).

Redefining Progress. 2006. Research institute studying practical ways to evaluate progress toward sustainability. http://www.rprogress.org (accessed 13 September 2006).

Robb Report Luxury Portal. N.d. http://www.robbreport.com (accessed 13 September 2006).

Scitovsky, Tibor. 1976. *The joyless economy: The psychology of human satisfaction.* Oxford: Oxford University Press.

Steg, Linda. 2005. Car use: Lust and must. Instrumental, symbolic and affective motives for car use. *Transportation Research Part A: Policy and Practice. An International Journal* 39(2–3): 147–62. www.elsevier.com/locate/tra (accessed 13 September 2006).

Stutz, John. 2006. *The role of well-being in a great transition.* Boston: Tellus Institute.

TRB (Transportation Research Board of the National Academes). 1997. Toward a sustainable future: Addressing the long-term effects of motor vehicle transportation on climate and ecology? TRB Special Report 251. Washington: US National Academy of Science Research Council, Transport Research Board, Committee for a Study on Transportation and a Sustainable Environment. http://www.trb.org/news/blurb_detail.asp?id=2672 (accessed 13 September 2007).

Tranter, Paul J. 2004. Effective speeds: Car costs are slowing us down. University of New South Wales, Sydney, Australia, for the Australian Greenhouse Office. http://www.greenhouse.gov.au/tdm/publications/pubs/effectivespeeds.pdf (accessed 13 September 2006).

Urry, John. 2004. The 'system' of automobility. *Theory, Culture and Society* 21(4/5): 25–39.

Veblen, Thorstein. 1899. *Theory of the leisure class.* Basingstoke, UK: Macmillan.

Verhoef, Erik T. and van Wee, Bert. 2000. Car ownership and status: Implications for fuel efficiency policies from the viewpoint of theories of happiness and welfare economics. *European Journal of Transport and Infrastructure Research* 0(0): 41–56.

VTPI (Victoria Transport Policy Institute). 2006a. Basic access and basic mobility. TDM Encyclopedia Online. http://www.vtpi.org/tdm/tdm103.htm (accessed 13 September 2006).

VTPI (Victoria Transport Policy Institute). 2006b. TDM marketing: Information and encouragement programs. TDM Encyclopedia Online. http:// www.vtpi. org/tdm/tdm23.htm (accessed 13 September 2006).

Chapter 12

The Global Intensification of Motorization and Its Impacts on Urban Social Ecologies

George Martin

As the use of one mode of transport – the private car – is both intensified and globalized, so are its social, ecological, and environmental consequences. While environmental consequences (such as enhanced carbon emissions) have received ample attention and study (Committee on the Future of Personal Transport Vehicles in China 2003), its social ecological consequences (such as growing transport inequity) have not (Adams 1999). The focus on the general environment is understandable given transport's substantial contribution to global warming. However, this 'big picture' can obscure the negative impacts of specific inequalities within transport, inequalities rooted in the differential distribution of its rewards and risks. These rewards and risks are defined by both social (i.e., social class) and ecological (i.e., location) parameters. For example, the public health impacts of pollution from growing traffic congestion disproportionately afflict poorer people who live in areas proximate to major road interchanges, petroleum refineries, etc. (Bae 2004).

Intensified motorization – that is, growing vehicle production and consumption per capita – takes different forms around the world, as do its local consequences. Among nations the major car consumers are also the major producers, a group situated in more developed countries (MDCs). In this group, there is national variability with regard to levels of urban motorized transport, as between the US and Japan. Moreover, despite globalization of both car production and consumption, cities in less developed countries (LDCs) still retain lower (and also varied) levels of motorized transport. This chapter analyzes these differences and their social ecological ramifications.

Global Motorization

The motor vehicle was developed in Western Europe and North America, and at the beginning of the twentieth century, its production was about equally divided between the two continents (Freund and Martin 2000). However, by 1910, with Ford's application of the mass production assembly line, the US began a domination of world output that lasted until the 1950s. In 1950, the US was the site of a staggering 76 percent of the world's motor vehicle production, and 70 percent of its consumption. After mid-century, production mushroomed in Western Europe and Japan, and for much of the latter half of the century, the Big Three – Germany, Japan and the US – dominated. In 1989, these three nations accounted for 58 percent of world production, and 50 percent of world consumption. By 2005 their shares had declined to 44 and 41 percent respectively.

As production is growing, motorization is now becoming fully global, with the greatest growth taking place in Eastern Europe, and in LDCs (Table 12.1). China jumped from being the twenty-third largest producer in 1989, to eleventh largest in 2002, and fourth largest in 2005. The Big Three are now the Big Four. Other LDCs, including Brazil, India, South Africa and South Korea, have also recorded large production increases.

Related to its dominance of vehicle production over the last century, the US has also been the largest consumer. With per capita vehicle registrations over 40 percent higher than the average for MDCs, the US continues to be in a consumption class by itself (Table 12.2). However, its consumption growth has slowed considerably – it was only 5 percent between 1994 and 2004. Other countries' consumption growth parallels production growth: Eastern European countries and LDCs, especially China, are the leaders.

Thus, motorization is globalizing rapidly but unevenly. Production is shifting from MDCs, especially the Big Three producers, to LDCs. Within MDCs, production is growing strongly only in Eastern Europe. On the consumption side, the US has reached saturation level. Growth is modest in Western Europe and Japan, strong in Eastern Europe, and spectacular in China. As the next section shows,

Table 12.1   World Motor Vehicle Production, Total, and Distribution, 1994, 2005

	1994		*2005*		*1994–2005*
	(millions)	*percent*	*(millions)*	*percent*	*change*
World	49.4	100	65.8	100	+ 33%
MDCs	41.5	84	46.1	70	+ 11%
Big Three	27.2	55	28.4	44	+ 5%
E. Europe	1.7	3	3.5	5	+107%
LDCs	7.9	16	19.7	30	+148%
China	1.2	2	5.7	9	+378%
India	0.5	—	1.6	3	+237%

Source: Based on WMVD 1996; MVFF 2006.

Table 12.2   World Motor Vehicle Registrations, Total and Per 1,000 People, 1994, 2004

	1994		*2004*		*1994–2004*
	(millions)	*per 1K*	*(millions)*	*per 1K*	*per 1K change*
World	629	115	837	136	+ 18%
MDCs	517	472	641	537	+ 14%
E. Europe	42	101	79	241	+139%
US	195	749	231	790	+ 5%
LDCs	113	26	196	40	+ 54%
China	9	8	28	21	+163%
India	6	7	11	11	+ 57%

Source: Based on MVFF 2006; WMVD 1996.

intensified motorization, especially in the US, has led to a hyperautomobility and its associated urban and suburban sprawl.

Intensified Motorization in MDCs

Originally developed in the US and Western Europe, motorized transport has peaked in the former. After World War II, the US Federal government promoted motorization through the construction of a vast interstate highway system and the marginalization of alternative modes of transport, resulting in auto-centred transport systems in most metropolitan areas (Freund and Martin 1993). These systems became the mode of mobility for developing auto social formations, characterized by built environments of single-family dwellings accessed only by car through extensive road networks (Martin 2007).

The motor vehicle saturation in the US is underscored when registrations are compared to the number of licensed drivers rather than to the whole population. In 2004, the rate was 790 vehicles per 1,000 people, but 1,164 vehicles per 1,000 licensed drivers. Thus for every licensed driver in the US, there are nearly 1.2 vehicles (MVFF 2006). National consumption data show a wide array for registrations, ranging from a high of 790 vehicles per 1,000 people in the US (hyper-motorized) to 0.7 in Bangladesh (non-motorized). The highlights of this array are the substantial gaps between the US and other MDCs, and between MDCs and LDCs. The US is a hyper consumer even when compared to the two nations closest to it with regard to culture, history and geography, as well as to level of development. US vehicle consumption is 22 percent greater than Australia's, and 38 percent greater than Canada's.

The latest development in auto social formations in the US is a deepening and broadening of personal car use: a hyperautomobility of saturated car ownership, more daily car trips, for longer distances, with fewer occupants (Martin 1999). While the number of vehicles per household in the US increased by a modest 6 percent between 1983 and 1995, the number of trips increased by 56 percent, and their length by 15 percent and the number of vehicle occupants per trip declined by 9 percent (DT 1995; 2000).

Hyperautomobility is also demonstrated by changes in vehicle miles (VMT) or kilometres (VKT) travelled. A transition to hyperautomobility in US urban areas is indicated by data comparing the growth rates in per capita urban VMT (car use) with growth rates in vehicle registrations (car ownership). The VMT increase was 3.3 times higher than the increase in vehicle registrations in the 1990s, compared to twice as high in the 1980s, and 1.3 times as high in the 1970s.

Hyperautomobility flourishes in the newer Sunbelt cities of the US South (such as Atlanta, analyzed in this volume by Henderson), and the West. Los Angeles is the prototype for the expansive combination of single-family housing and hyperautomobility. Between 1970 and 1990, the Los Angeles population grew by 45 percent, while its developed land area expanded by 300 percent (Diamond and Noonan 1996). In the US as a whole, urbanized areas expanded in the 1990s at about twice the rate of their populations (DHUD 2001). The expansion of land development at multiples of associated population growth is a feature of motorized urban sprawl.

Sprawl has been defined as "a process of large-scale real estate development resulting in low-density, scattered, discontinuous car-dependent construction, usually on the periphery of declining older suburbs and shrinking city centers" (Hayden 2004, 7–8). Because mobility is auto dependent, road congestion becomes the norm for much of the day, not only during work-commute times. Predictably, the Los Angeles metropolitan area leads the US in congestion delays: in 2003, LA drivers spent 93 hours (the equivalent of nearly 4 days) stuck in traffic, an increase of 46 percent over 1982 (TTI 2005).

The motorized urban sprawl of the US is indicated by comparative data on motorization and population density. Urban densities in the US are one-third those in Western European cities, one-tenth those in high-income Asian cities such as Tokyo, and one-fourteenth those in low-income Asian cities such as Bangkok (Kenworthy and Townsend 2004). In the US, 88 percent of urban trips are taken in private motor vehicles – 1.8 times that in Western European cities, 2.1 times that in high-income Asian cities, and 2.4 times that in low-income Asian cities.

While hyperautomobility can be identified with the US, increasing automobile use is a trend in other MDCs as well. The Joint Research Centre of the European Commission recently published a comprehensive study of Europe's growing sprawl problem entitled *Urban sprawl in Europe: The ignored challenge* (EC 2006). After growing by 36 percent between 1970 and 1980, VMT increased in the UK by 51 percent between 1980 and 1990 (DEFRA 2001). While car use is increasing in MDCs, the gap between the US and other nations remains sizeable. In 1998, it was followed by Canada and Australia, both at 63 percent of the per capita VKT in the US, Canada and Australia were followed by European Union countries at 52 percent and Japan at 30 percent of the US level (Fulton 2004, cited in Himanen et al. 2004; Newman 2000). Transport in Europe remains more multi-modal than in the US; for example, the levels of walking and cycling are about five times higher than in the US (Pucher and Dijkstra 2003). In Japan, the level of public transit use is even higher. Of course, these aggregated comparisons conceal considerable variability within nations; thus, New York City's modal split is much closer to that of Paris than that of Los Angeles (Newman and Kenworthy 1999).

The transport infrastructures of US hyperautomobility are both social (e.g., policing) and material (e.g., roads). They interact with natural topographies and built environments to create the signature habitats of an auto social formation, including far-flung exurbs, corporate campuses, malls, gated communities and big-box stores. This social formation has become typical in the US. Its basic parameters are low density and high motorization. One calculation of the relation between housing density and car travel is that a reduction from 20 to 5 dwelling units per acre increases travel by about 40 percent (VTPI 2004).

Automobility and Urban Social Ecologies in MDCs and LDCs

This motorized urban sprawl rests on transport injustice. Because its costs of participation rule out meaningful use by the poorest persons, and its dominance degrades low-cost public transport, the hegemony of the personal car in the US feeds inequality and injustice. Poorer people get poorer transport (Bullard and Johnson 1997; see Trumper and Tomic's chapter on Chile). Others experience transport inequity as well. The technical requirements of driving, especially on congested roads, exclude children, many people with disabilities, and frail elderly.

The environmental and public health impacts of motorization are usually measured by their gross effects across populations. However, many impacts (e.g., air pollution) are distributed differentially among social categories within populations. Such distributions and their territorial dimensions comprise the stuff of social ecologies. This differential distribution tends to overlay and intensify long-standing social class and racial-ethnic divisions within societies (see Henderson's chapter), as motorization develops its own social ecology of risk across metropolitan areas. Poorer people breathe poorer air and are more likely to live near 'hot spots', where pollution levels are much higher than in other parts of a metropolitan region. Many hot spots are generated via refineries, major road interchanges, and dense traffic (Bae 2004, 362).

In addition, the severances created by broad and heavily trafficked roads in MDCs foster social isolation for the elderly, the disabled, and the very young. Commentators have noted that the car "allows the ultimate segregations in our culture – old from young, home from job and store, rich from poor and owner from renter" (Calthorpe 1991, 51), and "creates differences in lifestyle practices around time and space, excluding many temporary or permanent non-users from participating in a variety of activities, denying 'citizenship'" (Jain and Guiver

2001, 572). This social ecological deficit of built environments dominated by motorization has also been noted in the UK:

> Poor people, and disadvantaged communities, often get penalized twice. Not only do they have to live with fewer economic resources, they often – indeed almost always – live in environments which exact an additional toll on their well-being, through being unhealthier, less accessible, and literally more expensive places in which to survive. The poor are more likely to live on inner city estates where overcrowding, high traffic density and lack of amenities are more common. (Worpole 2000, 9)

According to the UK's National Travel Survey, about 20 percent of households without access to a car have difficulty getting to supermarkets and doctors (DfT 2005).

The individualized and privatized hyperautomobility of auto social formations atrophies public space and attenuates public life (Freund and Martin 2007). It may contribute to the atomization of individuals and families, adding to the decline of community in the contemporary US. The decline of foot traffic in motorized urban sprawl indicates a more privatized social life enclosed within vehicles and homes. In addition to enhancing privatization, motorization can degrade commons through heavy traffic flows – plazas, parks, and squares can be rendered socially useless. Putnam (2000) identified sprawl as one of three aspects of US society that may explain the apparent decline in civic engagement. An OECD (2001) study found falling rates of civic involvement in some developed nations, especially in the US and Australia, both of which have very high levels of automobility.

A pressing issue in the major cities of LDCs is massive change in their geographies, built environments, and social ecologies, driven in large part by the increasingly widespread use of personal cars. Because of their scale and pace, these changes have significant global as well as local implications. While substantial research has been directed to the environmental impacts of foreign investment and economic liberalization (Fan and Lu 2001; Olds 2001) and to the impacts of increasing consumption (Sklair 2002; Wilk 2002) in LDC cities, little conceptual and empirical work has examined the impacts of new automobile-based landscape patterns, especially as they relate to changes in urban form, increased social fragmentation, and inequality (see Trumper and Tomic's chapter for an analysis that addresses this deficiency).

During the latter half of the twentieth century, LDC cities faced continual pressure and change from industrialization and new technologies such as electricity and motor vehicles, massive in-migration, and the emergence of wealthy residential enclaves. In almost all cases, cities rapidly expanded into their previous hinterlands (Ford 1996), became surrounded by migrant populations in squatter settlements, and were bisected by industrial and transport land uses. Up to 50 percent of residents in LDC cities now live in squatter settlements, which frequently lack running water and sanitation facilities, and are often located on or near hazardous sites. Near the centre of such cities, socially exclusive, multi-functional enclaves or 'citadels' emerge (Marcuse 1997).

On the heels of this production-oriented urban development, beginning in the 1980s and accelerating in the 1990s, new consumption patterns have been altering LDC cities. Beyond traditional city boundaries, unchecked new housing and transport development is creating a landscape that has many of the characteristics of sprawl. It is growth characterized by low-density and patchy development.

In some LDCs, government policy encourages the relocation of poorer inhabitants from older central residential areas to the periphery, thereby reducing the population densities in urban cores (Zhou and Ma 2000). Such a policy of 'forced suburbanization' has occurred in many Chinese cities as urban developers have cleared land in the centre for construction of new residential and commercial buildings. This process is similar to the construction of downtowns in US cities during the latter nineteenth and early twentieth centuries (Fogelson 2001).

Consensus is growing that transport is a critical ingredient in the mounting social ecological problems of LDC cities, as a study by the German government concluded concerning poverty:

> Further disadvantages arise from insufficient or lacking connections of poor settlements to public transport, considerable health hazards owing to settlements being built along roads or on the periphery of urban districts causing high levels of environmental pollution and high individual costs arising from transport expenditures. In the cities of many developing countries, poor families sometimes spend up to 20 percent of their income on transport, while the average family does not even require half that sum for its mobility needs. This stifles attempts to attain better living standards, which is why improving urban mobility represents an important step in combating poverty. (GTZ 2002, iv)

In LDCs, the motorization of transport systems currently falls into three general categories (Hu 2002; 2003; Kenworthy and Laube 1999; Vasconcellos 2001): Moderate, with about one-third the per capita VKT of the US, as in Latin America (e.g., São Paulo) and Asia (e.g., Seoul), and with substantial public transport; low, with about one-fifth the US level, as in Asia (e.g., Jakarta) and Africa (e.g., Cairo); and very low, with about one-tenth the US level, as in Asia (e.g., Beijing) and Africa (e.g., Abidjan), and with substantial bicycle, motorcycle, foot and animal transport, and some public transit.

While sprawl is not new in LDC cities, it is attracting new interest. For example, urban sprawl in China may have serious consequences because of the consequent loss of valuable agricultural land located in its coastal urban corridor. The land required for urban sprawl is now largely in farms. China's coast and major river valleys are home to both its most productive agriculture and largest urban concentrations, which get a large share of their food supply from local farmers. In 2003 alone, more than 2 percent of its farmland was lost to urban expansion (Yardley 2004). If China were to reach the car consumption levels of Japan, and "assuming the same paved area per vehicle in China as in Europe and Japan, a fleet of 640 million cars would require paving nearly 13 million hectares – most of which would likely be cropland" (Brown 2003, 50). This area would equal almost half of China's present 28 million hectares of rice land and would intensify the loss of cropland to settlement and infrastructure – between 1957 and 1990 the loss was put at 35 million hectares (Smil 1993; cited in McNeill 2000, 215).

Shanghai and other Chinese cities are sprawling. Decentralizing the core populations of cities makes room for the development of office buildings, tourist facilities and middle-class housing. The rate of growth of urbanized land in China between 1990 and 1995 was more than four times the rate of growth of its urban population (Zhang 2000). In the same period, the population density in the built-up areas of six Chinese cities declined 11 percent in just five years from 164 to 146 persons per hectare (Kenworthy and Hu 2002, 5). Authorities project Shanghai to expand to 1,100 km2 by 2010 (Mei et al. 1998; Wu 1999, cited in Committee on the Future of Personal Transport Vehicles in China 2003, 225), an increase of over 150 percent in less than a decade.

Growing sprawl enhances the social ecological impacts of motorization in China (Martin 2005). In Shanghai, social cohesion is reduced by rapid growth and the displacement of poorer residents: "With this dispersal of poorer communities to the periphery and the inclusion of high-cost housing in the mixed development in the central area, an increasing spatial polarization is taking place" (Newman and Thornley 2005, 242). A large in-migration of guest workers has intensified a recent rise in income inequality (Wu 2001). A recent report on China's 150–200 million rural-to-urban migrant workers pointed out their "poor to appalling" living conditions and their tendency "to live in highly concentrated communities, often on the outskirts of cities where there is little or inadequate infrastructure" (AI 2007, 27). Because China's development strategy promotes expensive car mobility to the detriment of inexpensive bicycle mobility (Committee on the Future of Personal Transport Vehicles in China 2003), transport is a key feature of the migrants' plight.

Large and rapid in-migration fosters an overlay of social inequalities with geographic locations for the poorer guests as opposed to the richer natives, especially when there are cultural differences (e.g., dialect) as well. In-migration has resulted in heightened social fragmentation and segregation in Chinese cities (Friedmann 2005) that is compounded by motorization. Motorization, while it may exacerbate segregation, does not create it. Thus, while the provision of cheap public transport may enhance the lives of dispersed (migrant) workers, it will not by itself decrease their spatial segregation.

Limits of and Resistance to Motorization

As Adams (1981, 47) pointed out over two decades ago, a motorized world built in the image of Southern California is not realistic on a global scale:

The American experience, desirable or undesirable, is not repeatable, because the conditions in which it occurred are not repeatable. The cities of the world that exist now and which were built before significant numbers of their inhabitants owned cars, have, with few exceptions, neither the undeveloped land, nor the fuel, nor the money necessary to follow the American example.

In addition to the resource constraints on the globalization of motorized urban sprawl, it faces growing popular resistance because of the limits to transport reform that rely only on technological fixes (see Wetmore, this volume, for a debate on social versus technological fixes). Cleaner engines and fuels reduce per car emissions but because of more car travel overall levels of pollution remain unacceptably high. Enhanced traffic management improves traffic flows, but these are subsequently overwhelmed by greater traffic volume. Finally, of course, few technological interventions are able to address the problems of land conversion, deterioration of commons and public transit, isolation and inequalities of social groups.

Historically, heightened motorization in MDC cities was grafted on to pre-existing multi-modal transport systems, and the focus of local politics has been on congested city centres and conflict over redevelopment, for example, turning brown fields into car parks or community gardens. Thus, while the North manages environmental degradation such as air pollution, it is faced with a growing buffet of issues flowing from contested right-of-way and land use claims among walkers, cyclists, gardeners, environmentalists, developers and motorists.

As motorized urban sprawl has become a dominant motif of US cities, its impacts have made it a focus of public concern and political controversy (Duany et al. 2000; Henderson, this volume). One reaction, based in the middle classes, focuses on quality of life issues and advocates regulating sprawl through such schemes as smart growth and transit-oriented development (Lehrer and Milgram 1996). Another, based in the working classes and the poor, focuses on social justice issues and advocates enhanced public transit (see Litman's chapter that draws on a social welfare perspective to argue for public transport policies). In the 1990s, new groups opposed to motorization blossomed in European and North American cities, including Critical Mass and Reclaim the Streets. Some linked transport directly to class and race; for example, the Bus Riders Union of the Labor/Community Strategy Center in Los Angeles (Bullard and Johnson 1997; Martin 2002).

In the South, access to transport for a growing poor and migrant labour population that is not motorized has emerged as a significant social ecological issue. For example, the Chinese government's restriction of bicycle use in favour of the car "is generating an intense debate between those attempting to facilitate traffic and those defending bicycles as effective and environmentally benign transport, especially for lower-income people" (Committee on the Future of Personal Transport Vehicles in China 2003, 143). Local resistance is also developing against road building:

Road building in cities is expensive and politically sensitive because of
difficulties encountered in acquiring land, relocating businesses and households,
displacing utility pipes and wires, and dealing with the confined conditions of
construction. Neighborhood interests also may resist construction – examples of
such resistance are already evident in Chinese cities. (Committee on the Future
of Personal Transport Vehicles in China 2003, 142)

Thus, even in China's relatively closed political system, dissent is growing
against the regime's all-out push for motorization as a prime driver of economic
development.

Conclusion

The expanded production of the auto through a complex of auto-oil-construction
firms is as fundamental a force behind the globalization of consumption landscapes
as are the telecommunications, fast food and computer 'revolutions'. This auto
nexus, coupled with the support of governments, is the basis of a powerful bloc
in the global economy. Automobility has become a determinant architectural,
geographical, and environmental influence that affects social ecologies in unique
and powerful ways – segregating classes, severing neighbourhoods, privatizing
public space, creating massive scales of operation, and leaving large ecological
footprints (Martin 2002).

While processes of population decentralization, geographical sprawl and
motorization are changing the social ecologies of many cities, the results are varied.
In the US, mass suburbanization is associated with lower population densities and
hyperautomobility, while European and Japanese cities have retained more of the
middle classes in their centres. Their suburbs, such as the *banlieues* of Paris, are
more densely populated, less auto dependent, and more working class than those
in the US. The development of suburban areas in LDC cities has been driven by
the need to locate housing for the mass influx of very poor migrant workers. Many
of the suburbs are densely populated but neither auto nor public transit dependent;
rather, they rely on very cheap, ad hoc forms of transport, primarily foot, cycle,
motorcycle, and 'jitney'. Private motorization is largely restricted to the elites and
middle classes who reside in centres.

While the problems associated with the social ecology of motorization are
substantial in the cities of MDCs, they pose greater difficulties in LDCs due to
their lower levels of development and regulation, and their faster pace of change.
The gradual changes in transport and housing that altered MDCs from the mid-
nineteenth century onward are being telescoped now in LDCs into just a few
decades.

As capital investment extends from the realm of production to that of
consumption, especially for personal housing and transport, income inequality
and social divisions grow. The potential for social fragmentation in LDC cities is

a central factor in their changing social ecologies. In this growing fragmentation, managerial and administrative employees connected to the import-export economy are the base for a new middle class. These higher income urban professionals increasingly seek private housing, private transport and private security near city centres. As a result, gated communities or 'security enclaves' are emerging rapidly in LDC cities, sharpening divisions between social groups.

Vasconcellos (2001, 158) uses the concept of an individualized motorization threshold to describe the current global situation and to recommend policy directions:

> Non-motorized cities, where pedestrian and bicycles are dominant modes, have a special need to protect and enhance such modes, while minimizing the possible negative impacts of motorization. Lightly-motorized cities face the challenge of how to accommodate increasing motorcycle use or rather decrease it. Heavily public-motorized cities face the challenge of improving accessibility and environmental conditions and avoiding transferring trips to private transport. Heavily automobilized cities face an immense challenge, that of minimizing environmental and inequity conditions that become extreme.

Ultimately, a new hybrid urban form is possible in LDC cities, emerging from the combination of rapid growth and globalization interacting with local ecologies and politics. The form could be a variation of the traditional LDC city that mitigates transport inequalities. It might include a built environment centred on a skyscraper Central Business District that serves as the locale for offices and residences of the elite and middle class. The sprawl outside this centre will be more extensive than it is today but may be managed through government regulation of high- and medium-density housing and transport. Near the outer borders of the metropolitan area will be satellite sub-centres and districts that contain production and assembly plants, infrastructure facilities, and housing for the working classes, especially migrant labourers, served by a dense network of public transport.

In considering new urban forms, the self-fulfilling or path-determining aspect to transport infrastructure is key. Once an auto-centred system is built, alternatives to the car become unusable (as Dennis and Urry in the next chapter put it, the system is locked-in). Subsequent reformatting of such a socially and materially embedded infrastructure is time consuming, capital intensive, and highly politicized. Contemporary US experiences in trying to stem and to retro-fit motorized urban sprawl in piecemeal fashion provide examples (see Henderson's chapter on Atlanta).

Thus, the complex nature of automobility places a high premium on anticipating transport systems in advance of land development. Perhaps this is the most important lesson that transport planners in LDCs can learn from the century-old motorization experience of MDCs. A 'soft' motorization that stresses small vehicle size and alternative fuels, combined with multi-modal and inter-modal transport infrastructures larded with cooperative uses of vehicles such as car-sharing schemes can create a path to social ecological justice in transport (see Dennis and Urry, this volume, for analysis of current prospects for a shift away from motorized urban sprawl). Democratic access to all spheres of daily life is more important now than ever, in the face of the increasing globalization of consumption landscapes.

Acknowledgement

The author acknowledges the support of the Centre for Environmental Strategy, University of Surrey, Guildford, UK, and the Department of Sociology, University of California, Santa Cruz, US.

References

Adams, John. 1981. *Transport planning: Vision and practice.* London: Routledge.

——. 1999. The social implications of hypermobility. In *Report for project on environmentally sustainable transport,* pp. 95–153. Paris: Organisation for Economic Co-operation and Development.

AI. 2007. China: Internal migrants: Discrimination and abuse: The human cost of an economic 'miracle'. Amnesty International Report. http://web.amnesty.org (accessed 01 February 2007).

Bae, Chang-Hee Christine. 2004. Transportation and the environment. In *The geography of urban transportation,* Susan Hanson and Genevieve Giuliano, eds, pp. 356–81. New York: Guilford.

Brown, Lester R. 2003. *Plan b.* New York: W.W. Norton.

Bullard, Robert D. and Johnson, Glenn S. eds. 1997. *Just transportation: Dismantling race and class barriers to mobility.* Gabriola Island, BC: New Society Publishers.

Calthorpe, Peter. 1991. The post-suburban metropolis. *Whole Earth Review* 78 winter: 44–51.

Committee on the Future of Personal Transport Vehicles in China. 2003. *Personal cars and China.* National Research Council, National Academy of Engineering, Chinese Academy of Engineering. Washington: National Academies Press.

DEFRA. 2001. *The environment in your pocket.* London: Department for Environment, Food and Rural Affairs.

DfT. 2005. *Focus on personal travel.* London: Department for Transport.

DHUD. 2001. *The state of the cities 2000.* Washington: Department of Housing and Urban Development.

Diamond, Henry L. and Noonan, Patrick F. 1996. *Land use in America.* Washington: Island Press.

DT. 1995. *Nationwide personal transportation survey.* Washington: Department of Transportation.

———. 2000. *Highway statistics series.* Washington: Department of Transportation.

Duany, Andres, Plater-Zyberk, Elizabeth and Speck, Jeff. 2000. *Suburban nation: The rise of sprawl and the decline of the American dream.* New York: North Point Press.

EC. 2006. *Urban sprawl in Europe: The ignored challenge.* Copenhagen: European Environment Agency, European Commission.

Fan, C. Cindy and Lu, Jiantao 2001. Foreign direct investment, locational factors and labor mobility in China, 1985–1997. *Asian Geographer* 20: 79–89.

Fogelson, Robert M. 2001. *Downtown: Its rise and fall, 1850–1950.* New Haven: Yale University Press.

Ford, Larry R. 1996. A new and improved model of Latin American city structure. *Geographical Review* 86: 437–40.

Freund, Peter and Martin, George. 1993. *The ecology of the automobile.* Montreal: Black Rose Books.

———. 2000. Driving south: The globalization of auto consumption and its social organization of space. *Capitalism Nature Socialism* 11: 51–71.

———. 2007. Hyperautomobility, the social organization of space, and health. *Mobilities* 2: 37–49.

Friedmann, John. 2005. *China's urban transition.* Minneapolis: University of Minnesota Press.

Fulton, Lewis M. 2004. Current trends and sustainability scenarios in transport energy use: Sustainable transport in Europe and links and liaisons with America. Paper presented in STELLA Focus Group 4 Meeting in Brussels, 25–27 March.

GTZ. 2002. *Urban transport and poverty in developing countries.* Eschborn: Division 44, Environmental Management, Water, Energy, Transport.

Hayden, Dolores. 2004. *A field guide to sprawl.* New York: W.W. Norton.

Himanen, Veli, Lee-Gosselin, Martin and Perrels, Adriaan. 2004. Impacts of transport on sustainability: Towards an integrated transatlantic evidence base. *Transport Reviews* 24: 691–705.

Hu, Gang. 2002. Land use and transport in Chinese cities: An international comparative study. Ph.D. thesis, University of Melbourne.

———. 2003. Transport and land use in Chinese cities: International comparisons. In *Making urban transport sustainable,* Nicholas Low and Brenda Gleeson, eds, pp. 184–200. London: Macmillan.

Jain, Juliet and Guiver, Jo. 2001. Turning the car inside out: Transport, equity and environment. *Social Policy and Administration* 35: 569–86.

Kenworthy, Jeffrey R. and Hu, Gang. 2002. Transport and urban form in Chinese cities: An international comparative and policy perspective with implications for sustainable urban transport in China. *DISP* [Zurich] 151: 4–14.

Kenworthy, Jeffrey R. and Laube, F.B. 1999. *An international sourcebook of automobile dependence in cities.* Boulder: University Press of Colorado.

Kenworthy, Jeffrey R. and Townsend, Craig. 2004. An international comparative perspective on motorisation in urban China: Problems and prospects. *International Association Traffic Safety Science* 26(2): 99–109.

Lehrer, Ute Angelika and Milgram, Richard. 1996. New (sub)urbanism: Countersprawl or repackaging the product. *Capitalism Nature Socialism* 7: 49–64.

Marcuse, Peter. 1997. The enclave, the citadel, and the ghetto: What has changed in the post-fordist city. *Urban Affairs Review* 33: 228–64.

Martin, George. 1999. *Hyperautomobility and its sociomaterial impacts.* Working Paper Series. Guildford: Centre for Environmental Strategy, University of Surrey.

———. 2002. Grounding social ecology: Landspace, settlement, and right of way. *Capitalism Nature Socialism* 13: 3–30.

———. 2005. The global diffusion of motorized urban sprawl: Implications for China and Shanghai. In *Urban dimensions of environmental change*, Huan Feng, Lizhong Yu, and William Solecki, eds, pp. 122–29. Monmouth, NJ: Science Press.

———. 2007: Global motorization, social ecology and the China question. *Area* 39(March): 66–73.

McNeill, John R. 2000. *An environmental history of the twentieth-century world: Something new under the sun.* New York: W.W. Norton.

Mei, Anxin, Wu, Jianping and Zhi, Luxi. 1998. Shanghai's land use. In *The dragon's head: Shanghai, China's emerging megacity,* Harold D. Foster, David Chuenyan Lai, and Naisheng Zhou, eds, pp. 119–39. Victoria, Canada: Western Geographical Press.

MVFF. 2006. *Motor vehicle facts and figures.* Southfield, MI: Ward's Communications.

Newman, Peter. 2000. *Sustainable transport and global cities.* Perth: Institute for Sustainability and Technology Policy, Murdoch University.

Newman, Peter and Kenworthy, Jeffrey. 1999. *Sustainability and cities: Overcoming automobile dependence.* Washington: Island Press.

Newman, Peter and Thornley, Andy. 2005. *Planning world cities: Globalisation and world politics.* London: Palgrave Macmillan.

OECD. 2001. *The well being of nations: The role of human and social capital.* Paris: Organisation for Economic Cooperation and Development.

Olds, Kris. 2001. *Globalisation and urban change.* Oxford: Oxford University Press.

Pucher, John and Dijkstra, Lewis. 2003. Promoting safe walking and cycling to improve public health: Lessons from the Netherlands and Germany. *American Journal of Public Health* 93: 1509–16.

Putnam, Ronald D. 2000. *Bowling alone: The collapse and revival of American community.* New York: Simon and Schuster.

Sklair, Leslie. 2002. *Globalization, capitalism and its alternatives*. New York: Oxford University Press.

Smil, Vaclav. 1993. *China's environmental crisis*. London: M.E. Sharpe.

TTI. 2005. *Roadway congestion in major urban areas*. College Station: Texas Transportation Institute, Texas A & M University.

Vasconcellos, Eduardo A. 2001. *Urban transport, environment and equity: The case for developing countries*. London: Earthscan.

VTPI. 2004. *Land use impacts on transport: How land use patterns affect travel behavior*. Victoria, Canada: Victoria Transport Policy Institute.

Wilk, Richard. 2002. Consumption, human needs and global environmental change. *Global Environmental Change* 12: 5–13.

WMVD. 1996. *World motor vehicle data*. Detroit: American Automobile Manufacturers Association.

Worpole, Ken. 2000. *In our backyard: The social promise of environmentalism*. London: Green Alliance.

Wu, Fulong. 2001. Housing provision under globalisation: A case study of Shanghai. *Environment and Planning A* 33: 1741–64.

Wu, Weiping. 1999. City profile: Shanghai. *Cities* 16: 207–16.

Yardley, Jim. 2004. China races to reverse its falling production of grain. *The New York Times* 2 May: 8.

Zhang, Tingwei. 2000. Land market forces and government's role in sprawl. *Cities* 20: 265–78.

Zhou, Yixing and Ma, Laurence J.C. 2000. Economic restructuring and suburbanization in China. *Urban Geography* 21: 205–36.

Chapter 13
Post-Car Mobilities

Kingsley Dennis and John Urry

There is no one ideal mode or speed: human purpose should govern the choice of the means of transportation. That is why we need a better transportation system, not just more highways. (Mumford [1953] 1964, 180)

You never change things by fighting the existing reality. To change something, build a new model that makes the existing model obsolete. (Buckminster Fuller)

They paved paradise and put up a parking lot. (Mitchell 1970)

Viewed through the lens of complexity, automobility is a self-organizing, non-linear system. It both presupposes and calls into existence throughout the world an assemblage of cars, car-drivers, roads, petroleum supplies, and other novel objects and technologies. The system generates the preconditions for its own self-expansion, including its elements, processes, boundaries, and other structures, and last but not least the unity of the system itself (Luhmann 1990). In this chapter, we use a framework of complex systems theory to assess the current state of automobility, as one possible mobility-system, and to project its future (see Dennis and Urry 2009 for more details).

Complexity

The emerging complexity sciences examine how components of a system through their dynamic interaction 'spontaneously' develop collective properties or patterns that are not implicit, or at least not implicit in the same way, within individual components (Urry 2003). Complexity investigates emergent properties, or regularities of behaviour that transcend the ingredients that make them up. Complexity argues against reducing the whole to the parts. In so doing it transforms scientific understanding of far-from-equilibrium structures, irreversible times and non-Euclidean mobile spaces. It emphasizes the nature of strong interactions occurring between the parts of systems, often without a central hierarchical structure that 'governs' and produces outcomes. These outcomes are both uncertain and irreversible.

Time and space are not to be understood as the container of bodies that move along these dimensions (Capra 1996; Casti 1994; Prigogine 1997). Time and space are viewed as internal to the processes by which the physical and social worlds themselves operate, helping to constitute their very powers. A further consequence of this fluidity of time is that minor changes in the past can produce potentially huge effects in the present. Such small events are not 'forgotten'. Chaos theory in particular rejects the common-sense notion that only large changes in causes produce large changes in effects (Gleick 1999). Rather relationships between variables can be non-linear, with abrupt switches occurring; the same 'cause' can in specific circumstances produce quite different kinds of effects (Nicolis 1995).

Complexity sees systems as being 'on the edge of chaos', where the components are neither fully locked into place but yet do not dissolve into anarchy. Chaos is not complete anarchic randomness; an 'orderly disorder' is present within all such dynamic systems (Hayles 1999). A particular agent rarely produces a single and confined effect. Interventions or changes will tend to produce an array of possible effects across the system in question. Prigogine describes these system effects as "a world of irregular, chaotic motions" (1997, 155).

Thus complexity examines how components of a system through their interaction 'spontaneously' develop collective properties or patterns. If a system passes a particular threshold with minor changes in the controlling variables, switches may occur and the emergent properties turn over. Thus a liquid turns or tips into a gas, or small temperature changes turn into global heating (Byrne 1998; Lovelock 2006).

Complexity and the Car

The system of automobility stems from the path-dependent pattern laid down in the 1890s. Once economies and societies were 'locked in' to the 'steel-and-petroleum' car, massive increasing returns resulted for those producing and selling those cars and its associated infrastructure, products and services (Arthur 1994). At the same time social life was irreversibly locked in to the mode of mobility that automobility both generates and presupposes. This mode of mobility is neither socially necessary nor inevitable yet it seems almost impossible to break away. From relatively small causes an irreversible pattern was laid down that has ensured the preconditions for automobility's self-expansion over the past 'century of the car'.

Predicting traffic expansion and then providing for it through new road building became especially marked during the middle years of the last century (Cerny 1990, 190–4). The car was associated with utopian notions of progress, and still is (see Trumper and Tomic, this volume, on Chilean modernity). The car's unrelenting expansion and domination over other mobility systems came to be viewed as natural and inevitable; nothing, it was thought, should stand in the way of its modernizing path and its capacity to eliminate the constraints of time and space (Sachs 1992). Over the century this naturalization of the car and its increasingly extensive lock-in with multiple organizations necessary for its expansion was facilitated through new discourses, such that drivers had to be qualified and appropriately trained, and that pedestrians should behave correctly so as to be able to cross roads safely in spite of their increasingly monopolization by cars.

'Path dependence' is key to understanding how patterns of socio-technical development are locked in through increasing returns. This notion emphasizes the ordering of events or processes over time. Contra linear models, the temporal patterning in which events or processes occur influences the way that they eventually turn out (Mahoney 2000). Causation can flow from contingent minor events to hugely powerful general processes that through increasing returns get locked in over lengthy periods of time. 'History matters' in processes of path-dependent developments (North 1990).

Path dependence happens when "small chance events become magnified by positive feedback" which 'locks in' systems so that increasing returns or positive feedback result over time (Arthur, quoted in Waldrop 1994, 49). Relatively deterministic patterns of inertia reinforce established patterns through processes of positive feedback. This escalates change through a 'lock-in' that over time takes the system away from what we might imagine to be the point of 'equilibrium' and from what could have been optimal in 'efficiency' terms, such as a non-QWERTY keyboard or electric forms of powering cars (Motavalli 2000).

In the 1890s the three main sources of energy for propelling vehicles were petrol, steam and electric batteries, with the latter two being more 'efficient' (Motavalli 2000; Scharff 1991). Petroleum fuelled cars were established for small-scale, more or less accidental reasons, partly because a petrol fuelled vehicle was one of only two to complete a 'horseless carriage competition' in Chicago in 1896. This was a turning point that signalled the beginning in America of the large-scale production of the automobile. In 1895 almost all cars manufactured in the world were made by three firms: Benz in Germany and P & L and Peugeot in France. By 1900 the French factories were producing around 4,800 automobiles; Germany 800; Britain about 175; and the Americans roughly 4,000. In around 1904–05 the US production began to overtake the leader France, reaching a production of 44,000 by 1907 (Bardou et al. 1982). Ransom Olds, who had built his first steam-powered automobile in 1886, switched to internal combustion to compete with the European models, and began large-scale production-line manufacturing of automobiles at his Oldsmobile factory in 1902. This concept was then famously, and successfully, expanded upon by Henry Ford into the car assembly line around 1914.

The economic gains accrued from the assembly line, mass production at lower prices, helped to bring about an automobile consumer class in a marketplace that had hitherto been reserved for the wealthy. Now the automobile could become a commodity for the masses. Also, the success of assembly line production gave the US automotive industry a dominating market position that quickly spread worldwide. The early 1900s thus saw a shift in automobile production from Europe to the US. The development of petrol-fuelled internal combustion engines is also a story of US industrial dominance over Europe, and the world, in the twentieth century. The next step within this automotive 'assemblage' was for car manufacturers to start sharing car parts with one another, resulting in larger production volumes at lower costs. Once established, the petrol system got 'locked in' to a path-dependent market and the rest is history. The automotive industry has since retained its interrelated network system as part of its global strategy, similar to how a complex system operates.

Small causes occurring in a certain order at the end of the nineteenth century turned out to have awesome and irreversible consequences for the twentieth century. Soon an array of other industries, activities and interests came to mobilize around the petroleum-based car, further strengthening the path dependency. As North (1990, 99) writes: "Once a development path is set on a particular course, the network externalities, the learning process of organizations and the historically derived subjective modelling of the issues reinforce the course."

As a consequence, institutions matter a great deal in how systems develop, contributing to their long term irreversibility. The effects of the petroleum car over a century after its chance establishment show how difficult it is to reverse locked-in processes, as billions of agents co-evolve and adapt to form a system of interdependent agents and relations – a complex assemblage that 'constitutes' the car.

The Car as a Complex Assemblage

One billion cars were manufactured during the last century, with now over 600 million cars roaming the world. World car travel is predicted to triple between 1990 and 2050 (Hawken et al. 2002). Country after country is developing an 'automobile culture' with the most significant at present being China (see Martin, this volume). By 2030 there may be over one billion cars worldwide (Motavalli 2000).[1]

The railway had in the nineteenth century initiated new cultural emphases upon machine-speed, timetables, punctuality, clock time and public spaces. But the emergence of the car system transformed that concept of speed into one of convenience. The car system provided a way of transcending a public timetable, enabling car-drivers to develop their own scheduling of social life. The car thus became the basis of autonomy, a vehicle for expressing personal freedom, an escape from the mundane into individualized leisure pursuits.

The 'structure of auto space' (Freund and Martin 1993) forces people to orchestrate in complex and heterogeneous ways their mobilities and socialities across very significant distances. The urban environment, built during the latter half of the twentieth century for the convenience of the car, has 'unbundled' territorialities of home, work, business and leisure (see Soron, this volume). Members of families are split up as they live in distant places requiring complex travel to meet up intermittently. People inhabit congestion, traffic jams, temporal uncertainties and health-threatening city environments as a consequence of

---

1   A 1997 report claimed that in that year, over 600 million motor vehicles existed in the world. It went on to predict that if the existing trends continued, that number would double in the next 30 years. In 2030, we could see 1.2 billion cars. See globalwarming. enviroweb.org.

being encapsulated in a privatized, cocooned, moving capsule (Whitelegg 1997; Miller 2001). People also use the car as a "medium for physical separation and physical expression of racialized, anti-urban ideologies" in such cities as Atlanta, Georgia (Henderson, this volume) to secede from what they perceive to be urban problems.

However used, automobility is a system in which everyone is coerced into an intense flexibility. It forces people to juggle tiny fragments of time so as to deal with the temporal and spatial constraints that it itself generates. Automobility develops 'instantaneous' time to be managed in highly complex, heterogeneous, and uncertain ways, in an individualistic timetabling of fragments of time. We might thus see the car system as a Janus-faced creature, extending individuals into realms of freedom and flexibility, but also constraining them to live spatially-stretched and time-compressed lives.

Within this 'machinic complex' car travel has ushered into modernity new experiences of stress, tension, and frustration, resulting in sporadic scenes of criminally violent behaviour classified as road rage (Lupton 2002). The car is the 'iron cage' of modernity, motorized, moving and privatized. Automobility thus produces desires for flexibility that only the car system can satisfy. Yet in order to cope with the 'mass' adoption of individualized automobility, a systemic assemblage of artefacts and support was required and developed.

'Automobility' is thus a hybrid assemblage, of humans (drivers, passengers, pedestrians) as well as machines, roads, buildings, signs, and entire cultures of mobility with which it is intertwined (Sheller and Urry 2000). It is not the 'car' as such that is key, but the system of these fluid interconnections since: "a car is not a car because of its physicality but because systems of provision and categories of things are 'materialized' in a stable form" that then we might say possesses very distinct affordances (Slater 2001, 6). It is necessary to consider what stable form or 'system' automobility constitutes as it makes and remakes itself across the globe.

Henderson (this volume) notes the frequent appearance of the 'inevitability hypothesis' in automobility discourse. In contrast a key feature of complexity that we emphasize here is that nothing is fixed forever. Abbott maintains "the possibility for a pattern of actions to occur to put the key in the lock and make a major turning point occur" (2001, 257). Such non-linear outcomes are generated by systems moving across what Gladwell terms 'tipping points' (2000). Tipping points involve three notions: that events and phenomena are contagious, that little causes can have big effects, and that changes can happen dramatically at a moment when the system switches, rather than in a gradual linear way. Examples include the consumption of fax machines or mobile phones, when at a moment every office needs a fax machine or everyone needs a mobile to keep up. In this context wealth derives not from scarcity as in conventional economics but from required abundance (Gladwell 2000).

Thus the issue for the current car system is whether a tipping point may occur when suddenly the world turns its back on it. What might complexity say about such a possibility? It should be noted to start with that we have only considered one form of the car and especially with how the 'path-dependence' of the privately owned and mobilized 'steel-and-petroleum' car was established and 'locked' in. Also we should note that current thinking about what is to be done about global automobility is characterized by linear thinking: can existing cars be given a small technical fix to decrease fuel consumption or can existing modes of public transport be improved and take some business away from cars and roads?

If the current automobility system is to undergo a shift it will surely not be because of a single 'fix'; rather, it will be the outcome of an 'assemblage shift'. This is likely to result from changes occurring within the 'mix' of relations that have formed around the current car transport system.

Shifting the System – An Assemblage of 'Tipping Points'?

Any post-car futures will also involve the futures of lifestyles, cities, architectures, thinking and attitudes. As the former Mayor of Bogotá, Enrique Peñalosa, said "we cannot talk about urban transport until we know what kind of a city we want, and to talk about the kind of city we want, we have to know how we want to live."[2] To know how 'we' want to live is fundamentally a question of values and 'patterns' of action. Complexity theory suggests that a major challenge is how to understand the dynamic and often non-linear changes that shift systems into a different pattern. Also, are there various tipping points that could create a break with the current car system? Complexity shows that such a system shift could be influenced by various interdependent small transformations occurring in a certain order that might move, or tip, the system into a new path (Sheller and Urry 2000; Gladwell 2000). These transformations may also be seen as various 'turning points', each one contributing to the system as it nears a potential 'tipping point'. Decades ago Mumford realized that: "in transportation, unfortunately, the old-fashioned linear notion of progress prevails...the result is that we have actually crippled the motorcar, by placing on this single means of transportation the burden for every kind of travel" (Mumford [1953] 1964, 177). To break with the current car system (Adams 1995, terms 'business as usual'), we need to examine what may constitute these various 'turning points', or transformations, that may ultimately lead towards 'tipping' the system.

The current structure of the car system is remarkably stable and unchanging, and supported through a huge economic, social and technological maelstrom of vested interests, agents and interrelated flows. Well over a century old and increasingly anachronistic because of its majority dependence on oil-based combustion, the car system still seems able to 'drive' out competitors, such as feet, bikes, buses, and trains. The twentieth century saw many homes and garages increasingly full of electric goods – except for the oddly out-dated car. However, some technical-economic, policy, and social changes could be laying down the seeds of a new

---

2   See Institute for Transportation and Development Policy (ITDP). http://www.itdp. org/events.html (accessed 14 July 2007).

mobility for the rest of this century. If they develop in the next decade or so, then a turning point could be reached through their systemic temporal interdependencies. If they occur in the 'right order', which we can only know in retrospect, they could 'tip' towards forming a new mobility, the 'post-car system'.

These small changes include: new fuel systems, for example, hybrid cars powered by diesel and batteries, and hydrogen or methanol fuel cells; new materials for constructing 'car' bodies that will be many times lighter; 'smart-car' technologies and communications that are embedded within and coterminous with various forms of transport; the deprivatization of the car through extensive car-sharing, car clubs, and car-hire schemes; and multi-modal transport policy away from predict and provide models. We now examine these potential small changes.

New Fuel Systems

The path-dependent system of the internal combustion engine is principally fuelled by the primary energy sources of oil, natural gas, and coal to produce petrol/gasoline, diesel, and LPG (liquefied propane gas). In 2005 the worldwide transport sector had a dependency on oil at 98 percent; this represented approximately 50 percent of all global oil consumption, and about 20 percent of all energy consumption, and followed an average growth rate of more than 2 percent per annum (Pinchon 2006). Presently 80 percent of the global oil reserves are controlled by national companies (Pinchon 2006). Remaining fossil fuels should be used as capital rather than income in order to invest in future projects and alternative transport scenarios. The future is likely to be punctuated by turning points in fuel systems making small improvements to a failing system, rather than revamping it entirely.

Fuel alcohols (ethanol/methanol) can be produced from a range of crops, such as sugar cane, sugar beet, maize, barley, potatoes, cassava, sunflower, eucalyptus, etc. (Salameh 2006). Two countries that have developed substantial biofuel programmes are Brazil (ethanol from sugar cane), and Russia (methanol from eucalyptus). Also, Malaysia and Indonesia are using palm oil whilst India is examining the potential of using jatropha, the physic nut. The worldwide production of ethanol in 2005 has been estimated at 37 million tonnes, 80 percent of this being used as fuel (Pinchon 2006). The largest producer is Brazil at 37 percent, with North America at 36 percent, Asia at 15 percent, trailed by Europe at 2 percent (others are 15 percent). The production of ethanol for fuel grew 15 percent between 2000 and 2005, with an estimated worldwide ethanol fuel use in 2050–2100 to be in the region of 33 percent (Pinchon 2006). Thailand is currently building over a dozen ethanol-production plants that convert sugar cane and rice husks. China, too, is engaged in biofuel plant construction and has already established the world's largest biofuel ethanol facility at Jilin. By the end of the decade the two most likely major ethanol producers will be Brazil and the US. A Biofuels Research Advisory Council (BRAC) report concludes by suggesting that by 2030 the European Union should supply up to 25 percent of its transport fuel needs by clean and CO2-efficient biofuels. Although biofuel usage cuts down dependence upon foreign oil imports, and is a cleaner, more environmentally friendly material, significant problems remain. The main argument against biofuel production in developed industrialized nations is the unavailability of agricultural cropland, and the shift in using food crop resources for fuel crops which has already had a negative effect upon food prices, especially detrimental to developing countries. For example, for either the US or Europe to replace 10 percent of their present transport fuels with biofuel using today's technologies would require up to 40 percent of their cropland (Salameh 2006) – an unsustainable amount. Other problems include related environmental problems of deforestation and land degradation; decrease in biodiversity; possible water pollution; an increased use of intensive biofuel farming using crop-spraying that emits polluting levels of nitrous oxide. Furthermore, a move to homogenized forms of biofuel production and cropping patterns in developing countries could be detrimental to existing agricultural practices leading to an irreversible use of agricultural land. This would further endanger the availability of food crops in developing countries. The UN has also warned that up to 60 million indigenous people may soon become biofuel refugees, with tens of thousands of rural families already displaced from their land by soya/biofuel companies.[3] Therefore, the BRAC (2006) report on future European biofuel usage recommends securing safe and consistent biofuel imports as well as developing possible alternative biotechnology programs. Biofuels may be significant in increasing future energy securities, yet research is still needed to engineer biofuel that is not dependent upon plantations (such as microbial energy conversion).[4] Other alternatives include the much-publicized hydrogen economy (Rifkin 2003) and electric batteries.

Whilst hydrogen is an attractive option, currently the "costs of producing hydrogen from renewable energy sources are extraordinarily high and likely to remain so for decades" (Romm 2004, 3). Infrastructure is also a problem. Two key issues for infrastructure involve where the hydrogen is produced, and how it is stored on the hydrogen vehicle. It may be necessary to develop a whole new delivery infrastructure as hydrogen, being highly corrosive, has the potential to make brittle current gas pipelines, thus requiring a new infrastructure of pipelines to be established using suitable materials (a problem similar to that of ethanol transportation). Overall, "hydrogen fuel cell cars are unlikely to achieve significant market penetration in this country (US) by 2030" (Romm 2004, 115). Similarly, Heinberg notes that "we need a solution now, not decades from now" (2004, 129). In this respect, hydrogen seems a poor short-term strategy, as well as unproven for the long term.

Current research is seeking to produce cheaper, more efficient lithium-ion batteries that are faster to charge. The MIT's new lithium battery contains

---

3   For more information see: http://www.biofuelwatch.org.uk/index.php (accessed 24 November 2007).

4   See recent report, Microbial Energy Conversion (Buckley and Wall 2006).

manganese and nickel, which is cheaper than the regularly used cobalt, and is capable of making the battery recharge up to ten times faster (Trafton 2006). However, the process still needs to be made cheaper before it can be produced commercially.

A variation on the hybrid-electric vehicle (HEV) that is gaining favour is the 'plug-in hybrid' car, referred to as Plug-In Hybrid Electric Vehicles (PHEV). PHEVs are grid-connectable Hybrid Electric Vehicles that are said to typically consume 50–90 percent less of any kind of conventionally used fuel. The PHEV is the basis for a viable Vehicle-To-Grid (V2G), Distributed Generation (DG) vision, whereby the car can be plugged back into the grid when not in use not only to recharge but also to return some of its unused energy during peak demand, for which the owner can be reimbursed. In this way credit can be earned which also encourages non peak-time driving. This peer-to-peer grid connectivity mirrors the web-like infrastructure of the internet and allows for distributed participation. It is similar to Rifkin's peer-to-peer vision for a hydrogen economy. It is claimed that "Electric-drive vehicles, whether powered by batteries, fuel cells or gasoline hybrids, have within them the energy source and power electronics capable of producing the 60 Hz AC electricity that powers our homes and offices."[5]

These new fuels, however, simply 'fuel' an existing linear transportation paradigm. Whilst they may solve issues of fossil fuel dependency and sustain desires for autonomous mobility, they still largely front an industrial establishment that promotes consumerism and consumption. They offer longevity to an existing paradigm that is under threat, but do not offer what is so badly needed – an alternative paradigm to an unsustainable global sociality.

New Materials

It is almost unavoidable that an efficient car will no longer be made of steel. The weight ratio needs remodelling since at present the car "needs only one-sixth of its available power to cruise on the highway and severalfold less in the city. The result is a mismatch not unlike asking a three-hundred pound weightlifter to run marathons" (Hawken et al. 2002, 27). It is possible that a shift in materials will influence the adaptation of smaller, lighter, safer, and smarter cars that increasingly are assembled to fit into an urban-industrial assemblage of software-architectures. The new materials will greatly facilitate the current drive towards 'Intelligent Transport Systems'. Another benefit coming from the adaptation of new, lighter materials is that of moving towards recyclable cars.

The European Union has proposed a goal of recycling 85 percent of vehicle components and converting 5 percent into energy by 2006. According to a new European Union legislation, by the year 2015, 95 percent of vehicle components will have to be recycled. This goal may signal a move towards systemic thinking in terms of industrial manufacture and materials. Industrial and social flows may

5   See V2G, N.d. Vehicle to grid power.

be forced to act as in ecological flows – in self-reflexive and recyclable processes, rather than in wasteful, linear flows. Too much energy is being wasted within the car itself – through heat dissipation from the engine, exhaust, tires, and so on. Recycling of energy needs to become more prominent within energy/fuel systems.

Nano-engineered materials are also likely to become part of a future redesigned vehicle. State-of-the-art research claims to have developed polymers filled with nanoscopic holes that are 'nanoporous polymers' for absorbing and storing hydrogen at increased rates for hydrogen cell fuels (Graham-Rowe 2006). The worldwide car market is heavily researching vehicle materials that will provide light material composition without sacrificing safety. Research on the 'hypercar', which uses advanced polymer composite materials (Hawken et al. 2002), is one example where materials are at the forefront of design and fuel efficiency. Other technologies include aluminum and nanotechnology which may make possible carbon-based fibres 100 times stronger than steel and at one-sixth the weight (US Department of Transportation 1999). Also production of much smaller micro-cars (rather than 4-person family-sized cars) for crowded urban spaces may increase (Urry 2007). Examples of such micro-cars include the Mercedes Smart Car, the Nissan Hypermini, Nice's Mega City and the Reva G-Whiz.

Wetmore (this volume) has traced the historical development of automobile safety since the 1960s and shows the contestations between the state and car manufacturers in the design and implementation of seat belts to airbags. In Wetmore's discussion the issue revolved around who, or what, would take responsibility for such automobile safety measures: individual drivers or the technological fix? MacGregor (this volume) also addresses what he terms 'the safety race' and how the need for public safety consciousness leads him to advocate stronger state intervention. MacGregor views movement in car safety as relying "critically on the state as an intervening authority between industry and road users." Whilst we share some of these concerns we would dispute the over-reliance upon top-down hierarchical forms of coerced change. Using a complexity approach we would be more inclined to view such possible changes, in car safety as well as other areas, as a combination of dynamic interventions from multiple sources. As Wetmore (this volume) concluded, "responsibility for addressing automobile safety must be distributed widely to ensure a greater possibility of success." This approach suggests that post-car automobility development is not a linear model of responsibility or development but a systemic arrangement of shifts and movements in several areas.

It seems likely that new players will emerge in the transportation construction market, such as electronics software companies as they integrate their products into hybridized cars. With the increase in the use of sensors future automobiles may resemble computers with wheels rather than cars with chips. Such cars will cross-over into being driven more and more by software than by hardware, and adaptable learning autocars will either have the advantage or be a necessity in order to integrate into the software transportation assemblage/ nexus. The 'lock-in' of a future path-dependence may force through embedded technologies rather than individual choice. In this scenario a turning point may well occur that coerces individual mobility into an assemblage of 'smart' technologies that constrain movement whilst simultaneously facilitating communication flows.

'Smart-Car' Technologies and Communications

For most of the twentieth century the revolutions in communication technology took place separately from the physical means of transportation. However, the current trend is toward the re-embedding of information and communication technologies (ICT) into moving objects. At the same time as information is being digitized and so released from location, cars, roads and buildings are being increasingly rewired to send and receive digital information in newly re-configured intelligent transportation systems (ITS). These reconfigurations could represent an epochal shift as cars are reconstituted as a networked system rather than separate 'iron cages', as a potentially integrated nexus rather than a parallel series. This could produce a shift from the modern divided traffic flow to what Peters (2006) terms the organic flow in which all traffic participants are able to survive and co-exist, aided by new kinds of communications regulating the overall system as a whole. MacGregor (this volume) considers this shift to be eliminating the forms of independence that may have been associated with earlier notions of automobility. To some degree this is true, yet rather than immobilizing the car user we see a potential shift in post-car mobilities towards increased integration within an overall systemic network which may very well increase the driver's connectivity and communication to 'bodies' external to the car and the car driving experience.

An issue that is of concern to post-car mobilities is how changes in automobility in the more technologized 'global North' might influence developing countries as they manage their own dynamic automobile systems. Elsewhere in this volume MacGregor argues that strong government/state intervention is required for the diffusion of affordable intelligent automotive systems and other safety innovations. In our view this underestimates the power of dynamic market forces, as well as the complex entanglements of local networks within civil society.

It is possible that smart-car technologies may accelerate the merging of cars into 'virtual' territories that will allow them to be intercepted from external sources whereby material updates will be provided wirelessly from manufactures such as in-car software upgrades via car-to-dealer communications. Fundamental to ITS will be the development of telematics, which includes wireless technology, vehicle tracking, and navigation assistance, and car-to-car communications. Already "Audi, BMW, DaimlerChrysler, Fiat, Renault and Volkswagen have formed the Car-2-Car Communications Consortium to seek consensus on standards for dedicated short range communication (DSRC)" (Bell 2006, 148). Longer range communications will come in the form of satellite tracking, which in Europe will be provided by the Galileo system,[6] due to become operational in 2008 yet currently 2–3 years behind schedule. The Galileo system is based on a constellation of 30 satellites in constant communication with ground stations in order to provide information on vehicle location, real-time navigation, speed control and potentially pay-as-you-go cost tracking. As part of the assemblage we expect to see ground networks embedding vehicle transport into systemic communication infrastructures. Overall, this will include short range car-to-car communications merging with cellular and radio frequency identity (RFID) transponders interfacing with satellite and state transport data systems (Bell 2006).

Transport/social environments may increasingly converge into coded space such that external architectures become a form of 'software-sorted geographies' (Graham 2005):

> we treat embedded intelligence as an abundant resource, having almost zero marginal cost, that will be deployed throughout society. This goes hand in hand with the continued expansion and development of the online world as we already know it. It is this availability of a pervasive computational mesh that makes intelligent infrastructure systems possible. (Sharpe and Hodgson 2006, 9)

This nexus of 'intelligent' cars may shift some present transportation systems into a Westernized form of centralized control obsessed with security and stability in a global climate of insecurity and unpredictability. This may inevitably raise issues of privacy, surveillance and state control as forms of future mobility are increasingly structured around the movement of information.

Deprivatization

The pattern of 'public mobility', of the dominance of buses, trains, coaches and ships, is unlikely to be re-established. It has been irreversibly lost because of the self-expanding character of the car system that has produced and necessitated individualized mobility based upon instantaneous time, fragmentation and flexibility. However, significant moves are taking place to deprivatize cars through car-sharing, cooperative car clubs, and smart car-hire schemes (see inter alia Hawken et al. 2002; Motavalli 2000). Even by 2001 six hundred cities in Europe had developed car-sharing schemes involving 50,000 people (Cervero 2001). Prototype examples developed in La Rochelle (Liselec), in northern California, Berlin and Japan (Motavalli 2000). Oxford has the UK's first hire by the hour car club scheme named Avis CARvenience. There are various other car clubs operating in the UK such as CityCarClub, Car Plus and Carshare. Two US car sharing companies are Flexcar and Zipcar. Canada has such coops as Communauto in Montreal and Co-operative Auto Network in Vancouver. In certain cases this

---

6   See EU 2007.

involves smart-card technology to book and pay and also to pay fares on public transport.

These developments reflect the general shift in contemporary economies noted by Rifkin (2000) from ownership to access, as reflected by the delivery of many services on the internet. We could hypothesize the payment for 'access' to travel/ mobility services will supersede the owning of vehicles outright. One important consequence is that if cars are not domestically owned then the various coops or corporations that provide 'car services' would likely undertake both the short-term parking and especially the long-term disposal of 'dead' vehicles. The former would significantly reduce the scale of car parking needed since vehicles would be more 'on the road', while the latter would radically improve recycling rates (as demonstrated in Hawken et al. 2002). Overall it is possible to propose a shift from cars as owned and driven by individuals, to deprivatized vehicles owned either by cooperatives or corporations, and 'leased'. This change may itself be reflected within newly emerging transport policies.

Transport Policy

In car transport, there is a noticeable shift away from predict and provide models based upon increased mobility as a desirable good towards new road schemes increasingly embedded within the emerging practical discourses of digitization, sustainability, and security. New road schemes are also likely to be networked within other systems and routes of travel. One example is the EU project, the Trans-European Transport Networks (TEN-T):

> By 2020, TEN-T will include 89,500 km of roads and 94,000 km of railways, including around 20,000 km of high-speed rail lines suitable for speeds of at least 200 km/h. The inland waterway system will amount to 11,250 km, including 210 inland ports, whilst there are a further 294 seaports and some 66 airports. (EU 2005, 7)

'New realist' policies involve many organizations developing alternative mobilities through integrated public transport, better facilities for cyclists and pedestrians, advanced traffic management, better use of land use planning, real time information systems, and a wider analysis of how transport impacts upon the environment (Vigar 2002). In particular, 'Western' transport policies increasingly take note of alternative models of transport developed, for example, in Curitiba, Brazil. This model involves separating traffic types and establishing exclusive bus lanes on the city's predominant arteries. As a result the bus service operates in a safe, reliable and efficient manner, avoiding the hazards and delays inherent to mixed-traffic bus services. As well, development along these bus routes has densified. Over a thousand buses make 12,500 trips per day, serving 1.3 million passengers. These schemes aim to model mass transit more upon the flexibility and efficiency of car-ownership. Five different types of buses operate in Curitiba, including a new 'bi-articulated' bus, the largest in the world, capable of carrying 270 passengers on the outside high-capacity lanes. Similar transport innovations are under development in Guatemala City; Pune, India (launched December 2006); Bangkok; Santiago, Chile; and Lagos, Nigeria (April 2007) (Transport-Innovator 2007). Such developments may help to ease some of the inner-city heavy traffic as car drivers will be encouraged to use these modernized and efficient bus services alongside other multi-modal forms of city transport such as connecting trams and metro. The city of London, UK is already investing in a new fleet of energy-efficient buses, including several hydrogen powered buses. These developments, it is hoped, will foster increasingly efficient transport networks within high-density city areas and encourage increased use of multiple forms of interconnecting public transport.

New transport policies likely to come into effect in high-technology regions are gradually expected to take into account security issues requiring systems for identity recognition. Also included will be measures to tackle general crime such as theft; this may involve remote immobilization. Electronic Vehicle Identification (EVI) allows for the identity of a vehicle to be read and recorded remotely. This will be promoted as providing security against uninsured, unlicensed, and untaxed driving; thus ensuring that all road users are legitimate and protected, safeguarding against unregistered drivers. However, once these measures are accepted in the public domain they may pave the way for more draconian policies. Already in the UK the Association of Chief Police Officers and the insurance industry have expressed a desire for electronic identification that extends to drivers through biometric recognition systems (UK House of Commons 2004, 43). Such a system may involve in-vehicle biometrics that checks the driver before permitting car use, such as in deterring those with above acceptable blood alcohol levels. The UK Association of Chief Police Officers and the motor industry are also developing a technology that will allow for external third parties (such as the police or various rental agencies) to enforce remote immobilization of stationary vehicles in the event of criminal misuse and illegal behaviour (UK House of Commons 2004). This technology is an extension of current services such as General Motors 'OnStar' in-vehicle safety and security system which offers 24-hour customer care access and immediate connection to emergency assistance in the case of an accident. Similarly, GM has announced that starting in 2009 they will equip all new models with an updated version of OnStar that allows police to enforce remote immobilization of the vehicle.

Thus future transport policies will increasingly address the social implications of the car system upon communities, land use and urban architecture. Also, transport policies will have to examine the relation between road planning and broader global transportation networks. A 'post-car' future requires not so much the re-invention of the car as the re-configuration of patterns of life, of introducing sustainable practices that presume quite different residential, work, and leisure mobilities. Creating a better car it will be realized will not 'fix' the problem of unsustainable patterns of life.

Conclusion

The power of automobility is the consequence of its system characteristics. Unlike the bus or train system it is a way of life, an entire culture, as Miller (2001) establishes. It has redefined movement, pleasure and emotion in the contemporary world. Sheller emphasizes "the full power of automotive emotions that shape our bodies, homes and nations" (2004, 237; Gilroy 2001). While the automobile may represent a love affair that expresses the values of individualism, freedom, and democracy, its status is highly contestable (as other chapters in this volume show).

The car system possesses distinct characteristics: it changes and adapts as it spreads along the paths and roads of each society, moving from luxury, to household, to individual item; it draws in many aspects of its environment which are then reconstituted as components of its system; the car system became central to and locked in with the leading economic sectors and social patterns of twentieth century capitalism; it changes the environment or fitness landscape for all the other systems; it promotes convenience rather than speed; the car system is a key component in the shift from clock to instantaneous time; it seems to provide the solution to the problems of congestion that it itself generates; it is able to externalize dangers onto those outside the system as it provides enhanced security for those within it; it is central to the individualist, consumerist culture of contemporary capitalism. Yet the days of steel and petroleum automobility are surely numbered.

In the next few decades it will seem inconceivable that individualized mobility will be based upon the nineteenth century technologies of steel bodied cars and petroleum engines. A tipping point (or series of turning points) will occur during the twenty-first century when the steel and petroleum car system will finally be seen as a dinosaur (a bit like the Soviet empire, early PCs or immobile phones). When such a tipping point will occur, however, cannot be predicted. It cannot be read off from linear changes in existing firms, industries, practices and economies. Just as the internet and the mobile phone came from 'nowhere', so the tipping point towards the 'post car' will emerge unpredictably. It will probably arrive from a set of technologies or firms or governments that are currently not a centre of the car industry and culture, as with the Finnish toilet paper maker Nokia and the unexpected origins of the mobile phone.

What is certain is that the linkages and connectivity in the post-car system will be of a more cyclic and interdependent nature, which will come from a convergence between sustainable energy management and technology management. Current automobility was developed around the use of unconstrained energy and materials, with linear flows between markets and consumers. Any future car system will have to "connect structure fundamentally to cyclic, systemic flows at every level, in the service of intensity and quality of life" (Sharpe and Hodgson 2006, 32). A post-car mobility will not be so much transformed through radical re-designs of the unit of mobility, but through the flows of mobility and interconnecting networks. 'Smart-car' technologies and communications will be needed to facilitate this shift, as discussed earlier. It is innovations in 'smartness' that will be one of the preconditions for moving rapidly towards networked mobility systems and the post-car mobility scenario.

The necessitated individualized mobility based upon instantaneous time, spatial fragmentation, and coerced flexibility[7] is seen by us to be potentially shifting from a series, or sequential platform, into a nexus. This nexus may see future post-car mobilities constructed into complex assemblages of networked structures, both natural and digital, that will combine individualized and social components into complex interconnectivities. Whatever any new system will be like, it will substantially involve a focus upon individualized and flexible movement that automobility has brought into being during the 'century of the car'.

We have examined post-car mobilities, or life beyond the car, as an assemblage system shift, knowing how linear transitions rarely account for major social and lifestyle upheavals. In this way such future(s) contain their own unpredictability. Knowing when and from where that all important tipping point may occur remains elusive, yet increasingly probable – or rather necessary. Then will it become clear what Mumford meant decades before when he said that "the only cure for this disease is to rebuild the whole transportation network on a new model" (Mumford [1953] 1964, 10).

References

Abbott, Andrew. 2001. *Time matters*. Chicago: University of Chicago Press.

Adams, John. 1995. *Risk*. London: UCL Press.

Arthur, W. Brian. 1994. *Increasing returns and path dependence in the economy*. Ann Arbor: University of Michigan Press.

Bardou, Jean-Pierre, Chanaron, Jean-Jacques, Fridenson, Patrick, and Laux, James M. 1982. *The automobile revolution: The impact of an industry*. Chapel Hill: The University of North Carolina Press.

Bell, Michael G.H. 2006. Policy issues for the future intelligent road transport infrastructure: IEE Proceedings. *Intelligent Transport Systems* 153(2): 147–55.

BRAC. 2006. *Biofuels in the European Union: A vision for 2030 and beyond*. European Commission: Biofuels Research Advisory Council.

Buckley, Merry and Wall, Judy. 2006. Microbial Energy Conversion. http://www.asm.org/Academy/index.asp?bid=46674 (accessed 22 February 2007).

Byrne, David. 1998. *Complexity theory and the social sciences*. London: Routledge.

Capra, Frijtof. 1996. *The web of life*. London: Harper Collins.

Casti, John L. 1994. *Complexification*. London: Abacus.

Cerny, Philip G. 1990. *The changing architecture of politics*. London: Sage.

---

7   For much detail, see Urry 2007, chapter 6.

Cervero, Robert. 2001. Meeting mobility changes in an increasingly mobile world: An American perspective. Paris: Urban Mobilities Seminar, June.

Dennis, Kingsley and Urry, John. 2009. *After the car.* Cambridge: Polity.

EU (European Union). 2005. Trans-European Transport Network: TEN-T priority axes and projects 2005. European Commission.

——. 2007. Galileo: European Satellite Navigation System. Directorate-General Energy and Transport. http://ec.europa.eu/dgs/energy_transport/galileo/index_en.htm (accessed 21 July 2007).

Freund, Peter and Martin, George. 1993. *The ecology of the automobile.* Montreal and New York: Black Rose Books.

Gilroy, Paul. 2001. Driving while black. In *Car cultures*, Miller, Daniel, ed., pp. 81–104. Oxford: Berg.

Gladwell, Malcolm. 2000. *Tipping points: How little things can make a big difference.* Boston: Little, Brown and Company.

Gleick, James. 1999. *Faster: The acceleration of just about everything.* London: Little, Brown and Company.

globalwarming.enviroweb.org. 2001. http://globalwarming.enviroweb.org/ishappening/sources/sources_co2_facts3.html (accessed 26 June 2007).

Graham, Stephen D.N. 2005. Software-sorted geographies. *Progress in Human Geography* 29(5): 562–80.

Graham-Rowe, Duncan. 2006. 'Nanoporous' material gobbles up hydrogen fuel. http://www.newscientisttech.com/article/dn10466 (accessed 13 November 2006).

Hawken, Paul, Lovins, Amory, and Lovins, L. Hunter. 2002. *Natural capitalism: Creating the next industrial revolution.* London: Earthscan.

Hayles, N. Katherine. 1999. *How we became posthuman.* Chicago: University of Chicago Press.

Heinberg, Richard. 2004. *Power down: Options and actions for a post-carbon world.* London: Clairview.

Lovelock, James. 2006. *The revenge of gaia.* London: Allen Lane.

Luhmann, Niklas. 1990. *Essays on self-reference.* New York: Columbia University Press.

Lupton, Deborah. 2002. Road rage: Drivers' understandings and experiences. *Journal of Sociology* 38: 275–90.

Mitchell, Joni 1970. *Yellow taxi.* In *Ladies of the canyon* (music recording) May.

Mahoney, James. 2000. Path dependence in historical sociology. *Theory and Society* 29(4): 507–48.

Miller, Daniel, ed. 2001. *Car cultures.* Oxford: Berg.

Motavalli, Jim. 2000. *Forward drive.* San Francisco: Sierra Club.

Mumford, Lewis. [1953]1964. *The highway and the city.* London: Secker and Warburg.

Nicolis, Gregoire. 1995. *Introduction to non-linear science.* Cambridge: Cambridge University Press.

North, Douglass C. 1990. *Institutions, institutional change and economic performance*. Cambridge: Cambridge University Press.

Peters, Peter Frank. 2006. *Time, innovation and mobilities*. London: Routledge.

Pinchon, Phillippe. 2006. *Future energy sources for transport*. Brussels: Future Energy Sources for Transport.

Prigogine, Ilya. 1997. *The end of certainty*. New York: The Free Press.

Rifkin, Jeremy. 2000. *The age of access*. London: Penguin.

——. 2003. *The hydrogen economy: The creation of the worldwide energy web and the redistribution of power on earth*. London: Polity.

Romm, Joseph. J. 2004. *The hype about hydrogen: Fact and fiction in the race to save the climate*. Washington: Island Press.

Sachs, Wolfgang. 1992. *For love of the automobile*. California: University of California Press.

Salameh, Mamdouh. 2006. *Can biofuels pose a serious challenge to crude oil?* Spring Croft, Haslemere, Surrey, UK: Oil Market Consultancy Service.

Scharff, Virginia. 1991. *Taking the wheel: Women and the coming of the motor age. New York:* Free Press.

Sharpe, Bill and Hodgson, Tony. 2006. *Intelligent infrastructure futures technology forward look: Towards a cyber-urban ecology*. London: Office of Science and Technology.

Sheller, Mimi. 2004. Automotive emotions: Feeling the car. *Theory, Culture and Society* 21: 221–42.

Sheller, Mimi and Urry, John. 2000. The city and the car. *International Journal of Urban and Regional Research* 24: 737–57.

Slater, Don. 2001. Markets, materiality and the 'new economy'. Paper given to Geographies of New Economies Seminar, Birmingham, UK, October.

Trafton, Anne. 2006. MIT powers up new battery for hybrid cars. http://web.mit.edu/newsoffice/2006/battery-hybrid.html (accessed 24 October 2006).

Transport-Innovator 2007. *Transport innovator: January–February 2007*. Washington: Bus Rapid Transit Policy Center.

UK House of Commons. 2004. *Cars of the future*: Seventeenth Report of Session 2003–04, HC 319-I. House of Commons Transport Committee.

Urry, John. 2003. *Global complexity*. Cambridge: Polity.

——. 2007. *Mobilities*. London: Polity.

US Department of Transportation. 1999. Effective global transportation in the twenty-first century: A vision document. US Department of Transportation: 'One Dot' Working Group on Enabling Research.

V2G. N.d. Vehicle to grid power. Newark, DE: University of Deleware. http://www.udel.edu/V2G/ (accessed 27 July 2007).

Vigar, Geoff. 2002. *The politics of mobility*. London: Spon.

Waldrop, M. Mitchell. 1994. *Complexity*. London: Penguin.

Whitelegg, John. 1997. *Critical mass*. London: Pluto.

Index